F. L. Jenkner

Electric Pain Control

Springer-Verlag Wien GmbH

F. L. Jenkner, M.D., F.I.C.S., F.N.Y.C.S.

Wallrießgasse 26, A-1180 Vienna, Austria
A.o. Professor of Surgery/Neurosurgery,University of Vienna Medical School
Formerly: Head of the Neurosurgery and Pain Clinic of Ambulatorium Süd, Vienna
Former Pain Consultant to the "Ludwig Boltzmann Institute
of Clinical Oncology" at the municipal hospital "Lainz" of the City of Vienna
Guest Professor at Georgetown University, USA
and at University of Manitoba, Canada
Elected active member of The New York Academy of Sciences

Typesetting: Camera ready by author

Printed on acid-free and chlorine-free bleached paper

With 112 Figures

ISBN 978-3-211-82622-5   ISBN 978-3-7091-3447-4 (eBook)
DOI 10.1007/978-3-7091-3447-4

1 Capitis dolorē quemuis ueterem &
intolerabilem protinus tollit,& imper-
petuum remediat torpedo uiua nigra,
impofita eo loco q in dolore eſt, donec
deſinat dolor, & obſtupeſcat ea pars:
quod cum primū ſenſerit, remoueatur
remedium , ne ſenſus auferatur eius par
tis. Plures aūt parādæ ſunt eius generis
torpedines : quia nōnunꝗ uix ad duas,
trésue reſpondet curatio,id eſt, torpor,
quod ſignū eſt remediationis.

SCRIBONIUS LARGUS DESIGNATUS
"Scripta mea Latina medicalis, codex instar"
editio prima, cap. XI. De compositione medicamentum liber.
Ed. J.M.Berthold, Argentor, 1786.

CONTENTS

THOUGHTS ABOUT THE FRONTISPIECE

From Roman times, the most famous collection of prescriptions, also containing remedies for painful states, has been transmitted to us in original phraseology (even though the original writing has been lost). It had been written by the physician and writer SCRIBONIUS LARGUS DESIGNATUS between 43 and 48 A.D. for the personal use of C. Julius Calliostus who carried it with him during the campaign against Britannia.

SCRIBONIUS LARGUS was the son of a liberated slave ("libertus") and lived in Rome. Whether he also was born there is unknown. He was a scholar (= student) of CELSUS and TRYPHON during the reign of Tiberius (from 14 A.D.). One of his "comilitones" (we could say "class mates" to-day) was a certain VETIUS VALEUS, who had, during a certain time, Messalina (wife of emperor CLAUDIUS) as his concubine and was killed with her as reported by TACITUS (ann.,11:35). SCRIBONIUS LARGUS should have been born in the year 11 B.C., since it is reported that he celebrated his $\alpha\kappa\mu\epsilon$ (Greek for: at his prime, standing in full blossom), which meant his 40th birth day, in the year 29 A.D.; accordingly, he must have been 54 years old when he accompanied emperor CLAUDIUS to BRITANNIA and started to write the collection of prescriptions which made him famous up to our time. He was 59 years of age when he finished this work. He lived under the emperors TIBERIUS (14 – 37), CALIGULA (37 – 41) and CLAUDIUS (41 – 54). Some of these dates may be deduced from facts reported in his writings: e.g. he had described a certain plant (trifolium acutum, in chapter 163) which he reportedly saw when he was "with his god, caesar, visiting Britannia". Or he mentions a dentifrice (dentifricium; in Chapter 60) which "was used by Messalina".

His name actually means "the one writer, giving much" or the "poet...". He was not a member of the noble family of the "SCRIBONII" which is clearly stated by his agnomen "DESIGNATUS" (meaning "named as such"). About his position we only know that he did not reach the rank of an "imperial physician" of CLAUDIUS, since this position was held by SERTINIUS who obtained a reported salary of 600.000 sesterces (unit of currency). But he was a battle - or military-physician and during a certain time he was personal physician to Messalina (the empress). Other personal physicians to Roman emperors had been ANDROMACHUS for NERO; ANTONIUS MUSA for AUGUSTUS and ORIBASSUS for JULIANUS APOSTATA.

The face of SCRIBONIUS LARGUS we are able to grasp from a vase (made in 1624) on which portraits of the eight famous physicians from antiquity were depicted: ANDROMACHUS, ALI ABBAS, CELSUS, SCRIBONIUS LARGUS, HIPPOCRATES, GALEN, PAULUS OF AEGINA and AVICENNA. This vase had been described by Ed. BONNET as "une vase à thériaque et la confection de la Thériaque à Toulouse (1898)".

SCRIBONIUS LARGUS, the author of our frontispiece, had – for certain – written his prescriptions in Latin. Aside from his splendid way of writing there are to be found some new words which he initiated or phrases (such as: "ut ita dicam" = so to say) which a translator would never have been able to form, in case the original would have been written in Greek. However, there are several famous Roman writers, who wrote their texts in Greek: PLINIUS the elder, SEXTINIUS NIGER or JULIUS BASSUS. It is not entirely clear whether SCRIBONIUS also did write in Greek; the collection of prescriptions we report on here is headed by the remark

"scripta mea latina"

meaning "my Latin writings"; and when stressing the word "latina" there may be existing also "scripta graeca" by the same author.

The original hand writing was lost (the editio princeps) and as the original phrasing of the text there is regarded the writing of "CODICIS INSTAR" which is reported to be identical to "CODEX LAUDUNENSIUS" (420, from Lâon). Many citations from SCRIBONIUS LARGUS are known to exist in the writings of a large number of famous physicians (and not only Romans) living after him. GALEN especially was writing and citing our author. Whether there also existed a Greek text of the collection which may have been retranslated into Latin under the reign of emperor VALENTINIAN (4th century A.D.) is most improbable.

Aside from the well known work of the collection of prescriptions, SCRIBONIUS LARGUS had initiated a listing of official volumetric measures and weights of his time which reportedly were made uniform and transmitted to wide usage by him. These were

volumetric measures

Latin			Latin		metric		US units
CONGIUS	=	6	SEXTARII	=	3600 ml	=	3.8 quarts
SEXTARIUS	=	1/6	CONGIUS	=	600 ml	=	20 fl.oz
HEMINA	=	1/2	SEXTARIUS	=	300 ml	=	10 fl.oz
CYATHUS	=	1/12	SEXTARIUS	=	50 ml	=	1.7 fl.oz

weights

Latin			Latin		metric		US units
LIBRA	=	12	UNCIA	=	360 g	=	12.7 oz
SEXTANS	=	1/6	LIBRA	=	60 g	=	2 oz
UNCIA	=	7	DRACHMAS	=	26.25 g	=	1 oz
DENARIUS	=	1	DRACHMA	=	1.75 g	=	27 grains
(Roman)			(Greek)				
VICTORIANUS	=	1/2	DENARIUS	=	0.9 g	=	14 grains
SCRIPULUM	=	1	OBULUS	=	1.25 g	=	19 grains

From the number of copies, rewritings and reprintings of the collection of prescriptions of SCRIBONIUS LARGUS, with and without commentaries, one may surmise that this collection always was regarded as a very important piece of work in medical literature. The most important editions are: Paris 1529 (Ruellis), Basle 1529 (Ruellis), Basle 1547 (Aldus), Henricus Stephanus 1567, Padua 1655 (Rhodius), Berthold 1786 and G. Helmreich 1887. These dates are taken from the most exact studies of W. SCHONACK, Fischer, Jena, 1912, who also wrote a complete German translation with commentaries in 1913. A very elaborate commentary is the edition of Rhodius from Padua (1655).

All historians assume SCRIBONIUS to be neither a pure empiricist nor a dogmatic or methodologic writer. He was judged an ecclectic, a practical researcher. Therefore, the importance of his report on a subject of utmost importance for us is generally accepted and documented. The application of a widely existing fish, the

TORPEDO NIGRA = electric ray

as a remedy for pain touches on our very subject here. It also interests us for the quality of impulses emitted by this ray, as we shall discover soon. SCRIBONIUS LARGUS reported on the use of this fish in cases of headache (chapter XI) and gout (podagra; chapter CLXII). Our frontispiece is the reprint of chapter XI and the English translation by the author of this monograph is the following:

Chapter XI.: Ad capitis omnum dolorem, et quamvis veterem qua res in perpetuam sanat. ON HEADACHE, AS LONG AS IT MAY BE LASTING AND HOW TO TREAT IT PERMANENTLY.

Unbearable headaches, as enduring as they may be, are relieved immediately and cured forever by the living black electric ray if it is held in contact with the aching area until pain disappears and this part is numbed; as soon as this is felt, the remedy should be removed such as not to interfere with sensation. There are several species of these fishes; since the effect does not occur at once, but after 2 or 3 applications, healing, that is numbness which is a sign of healing ...

Chapter CLXII.: For every type of podagra place an electric ray under the feet, just when the pain comes on. But do not stand on a dry part of the beach, but where the sea is wetting it. Keep the contact to the ray long enough to make foot and skin areas numb up to the knee. Then the pain will disappear immediately and will hold cured for the future. In this fashion Antheros, a public employee, who had been (a) liberated (slave) by Tiberius, was cured.

In his commentary to chapter XI J. Rodius (Ioannis Rodii ad Scribonium Largum Emendationi, et notae, Patavii 1655, Typus Pauli Frambotti Bibliopola Superiorum permissu) had mentioned that an unknown author had pointed out that the electric ray cures very molesting headaches in old people (Narca marina antiquis passionibus apposita sedat vehementes dolores capitis). Clarius Dioscurides, Paul Aeginatus, Aetius Tetrab., Galenus, Math. de Gradi. Plinius, Plutarch, Petrus Bellonius (de aquatibus) and Julius Scaliger all had reported on the favorable effect of the electric ray in cases of headaches.

In the remarks to chapter CLXII one may find that Athenaeus, Paulus Aeginatus, Plinius, Marcellus, Annaeus and Aelius Praetorius had reported on pain from gout (podagra), just as Plinius also reported that M. Agrippa lost his pain in the leg in this manner.

The observation that TORPEDO NIGRA has an excellent effect on pain and is able to cure it has to be regarded as a strictly empirical one. This is so because at the time of SCRIBONIUS LARGUS one did not have any concept of the true nature of electricity. In using certain forms of electric current for therapeutic endeavors we are connecting certain concepts with it, concepts of the kind of impulses which may be useful. What now may the electric ray be able to do, as seen from our views? Is there possibly any connection to the kind of electric impulses it is giving off?

The fish TORPEDO NIGRA, as it was named by our author SCRIBONIUS LARGUS, is named to-day TORPEDO NOBILIANA and prevailed in the Mediterranean in the antiquity as well as it does to-day. It may become as large as 1.6 m (5.3 feet). Its electric organ actually is derived from a muscle and consists of small roundish sheets, one on top of the other, which conduct like a Volta's column: this organ may deliver DC-impulses from 40 to 220 V of a short duration. This actually represents the reason why we have selected the Latin citation as a frontispiece: JENKNER and SCHUHFRIED [130] could prove that among the various different electrical impulses examined for their effect on thin nerve fibers those impulses of DC having a very short duration of a single pulse have the optimal desirable effect, and this effect is reproducible. The difference to other forms of electric impulses is very significantly demonstrable by statistical methods ($>.01$) and the very short DC-pulses have an effect lasting the longest. We shall point to this difference in the course of this monograph and present it in an objective manner. We shall also draw conclusions essential for discussing methods of non-drug therapy for pain by electric nerve blocks, the central subject of this monograph [118].

5

I. General Section

INTRODUCTION

Cui placet, obliviscitur
cui dolet, meminit.
(Cicero; pro Murena, 42)

Pain problems represent difficult tasks for every physician, challenging his knowledge in the theoretical disciplines of anatomy and physiology. More than for other problems this information is essential for correct evaluation of causes for pain and selecting and achieving optimal pain therapy. To discuss pain problems has become very modern, but by far not all that is written is to the point or accurate. The problem does not become easier if patients take the drugs prescribed for them indiscriminately or more often than prescribed; on the other hand there are those physicians (who really are not physicians, I think) prescribing just about any analgesic or giving an injection for the sake of getting rid of the patient fast. Therapy of pain without drugs becomes more and more important every day. For pain control without drugs electric current always has played an important role. This has been so even at a time, as we know to-day, when men had not the slightest idea about the nature [118] of an electric current. In proof of this there is pictorial evidence where there are no written characters [GAILLARD; 72; 2750 bC] and written records [SCRIBONIUS LARGUS, 215]. When the nature of an electric current became clarified, the first application of such a current was a medical one [179]. Shortly thereafter a report on dental-medical use appeared. Towards the middle of the last century first attempts of systematic applications began [ALTHAUS, 4] using the electric current in treating various pain conditions. The discipline of physical medicine has standardized various applications and types of electric current, aside from other forms of treating pain without drugs, for various states: galvanic (direct) current, pulsed galvanisation, surging current, exponential current as well as other special types of current, e.g. interferential current, have been assigned to be used for special indications. On these "classic types" of current, we shall not report here, since there are existing enough text books [61, 109].

Recently two authors [MELZACK and WALL, 182] presented an entirely new (at the time of first publication) theory of pain. The origin and prevention of pain were understood in a different way giving rise to the use of special types of current and special types of applying it to a patient. These new types of electric impulses having been advanced not always in a correct manner, certainly not studied enough as to their effectiveness in relieving pain, have been reported to "...be regarded as unsatisfactory"[204] or "...should be studied more closely"[205]. Reports on an effective pain reduction of about between 30 and 40%, maximally 50%, appeared. But the very differing types of electric impulses were said to all have identical effects. This did not seem rational to us. Seeing that operative deductions of the "new" pain theory for dorsal column stimulations (DCS) brought only about 10% pain relief we regarded an experimental approach as most essential to clarify what effect any of these electrical impulses could have on nerve fibers if applied from the skin. If an effect was observable, what kind would this effect be. We could not agree that without knowing the type of action an electric field should have on nerves a terminology for this "new" therapy should be coined. Instead, we thought that applying short electric impulses to the skin (which could be called "stimulation") should not generally be called "stimulation" or more specific "Transdermal Electric Nerve Stimulation" unless one had proof that the effect of these impulses results [117] in stimulation. To select a terminology without having any evidence for it seemed at least rash. Therefore, we looked for a model to test whether thin nerve fibers would respond to any electric impulses and study what kind of effect such an electric field would have on these small diameter fibers. We found such a model and have reported on the studies [JENKNER and SCHUHFRIED, 130]. The results of these studies shall be mentioned and summarized together with conclusions to be drawn from these on a possible electric method of pain control under the heading of

9

"optimal wave form". First let us discuss a few important facts from neurophysiology which help us to better understand the origin of pain, pain sensation and the manner leading to reduction of this sensation. We therefore suggest to the reader not to skip the following brief paragraph on the basic theory behind our subject.

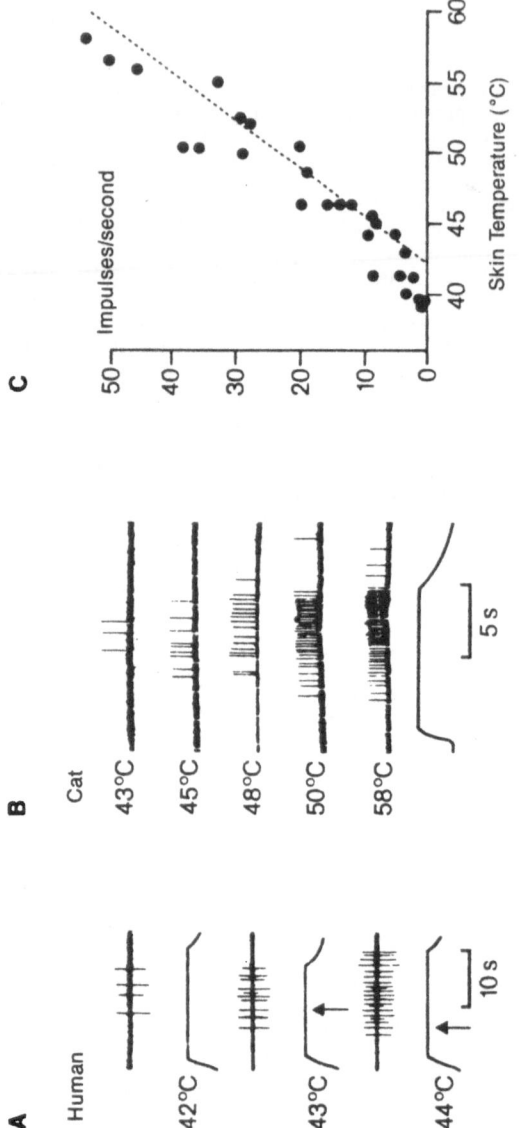

Fig. 1: Action potentials of afferent C-fiber of human radial nerve (A) and of posterior tibial nerve of anesthetized cat (B) in relation to skin temperature. Histogram (C) showing interrelationship of action potentials of cat to skin temperature. Arrows in (A) indicate subjective pain threshold of subject. From ZIMMERMANN [254] with kind permission of author and publisher, which relates also to fig. 2 and 3.

NEUROPHYSIOLOGICAL BASIS

Pain ist life (Schiller,
Wilhelm Tell IV, 2; Attighausen)

Nerve conduction in general is an electro-chemical [138] event; conduction of pain therefore also is of such a nature until it meets certain structures within the brain (basal ganglia). It may be modified or influenced not only by pharmacological (i.e. drugs) but also by electrical means. After the algophorus impulses meet certain cerebral areas, we sense "pain", just as certain other impulses turn to our sensing of light, sound, touch etc. in other areas of the brain. To influence our pain sensation would seem to be most effectively and very simply done by interrupting the pathways along which these electro-chemical events run centripetally. This applies especially if the cause of pain may not be visualized and may not be removed. The relative strength or degree of any sensation is made aware to us by a very special way of coding of these events. The main advantage of this system of coding by modulation of pulse density is that the information contained in a single spike (action potential) may vary by 500% and drop to as low as 20% of the original size without losing any information; information which is made available in the next proximal synapse, or the last end of the pathway within the cerebral cortex just as well, by decoding. The sensation of "weak pain" is caused by just a few "spikes" of action potentials running centripetally over the pathways, while "stronger pain" is indicated to us by a greater number of spikes per sec. arriving in the brain from the periphery. This may be demonstrated very impressively for the sensation of temperature (fig. 1). By this experiment one may determine objectively the strength of a sensation (or pain) or – expressed in a different way, if a sensation of pain is increasing (in this case, the frequency of spikes is increasing) or decreasing (then frequency of spikes would be decreasing). Applying this knowledge on the experiment depicted in fig. 2 one may notice that n- adrenaline at the free nerve endings will increase spike frequency just as will stimulating a sympathetic fiber. This, however, means that by these events pain sensation will increase in strength. The experiment presented in fig. 3 reveals: While radiant heat applied to the skin will increase spike frequency of a recording of a medullary C-neuron, simultaneous stimulation of A-fiber using 50 hz frequency of impulses will definitely decrease spike frequency: apparently simultaneous "electrostimulation" (or whatever you may call it) has a pain relieving effect. The limits of spike frequencies for coding are between 18 and 800 per second [139]. In the various sensory modalities the relation of strength of subjective sensation and of the strength of stimulus never is identical, rather a very different exponential function

$$E = k \cdot R^n$$

characterizes the single senses in a different way, as may be visualized in fig. 4 [STEVEN's exponential function, from KEIDEL, 139]. The exponent "n" is the greatest for pain and the smallest for vision. This has an important message: compared with other sensory modalities, for pain the smallest increase in strength of stimulus leads to a higher - than - threshold sensation; expressing it in other words, this indicates that pain has only few distinguishable increments of strength. The practical value of this characteristic is eminent; a patient who just started to sense a toothache will - on a very minimal increase of stimulus strength - sense maximal pain and see a dentist. Very differently for vision, a very large number (55!) of increments of stimulus strength is necessary to allow us to reproduce (i.e. perceive) our environement most accurately. These differences may also be observed objectively [KEIDEL, 139].

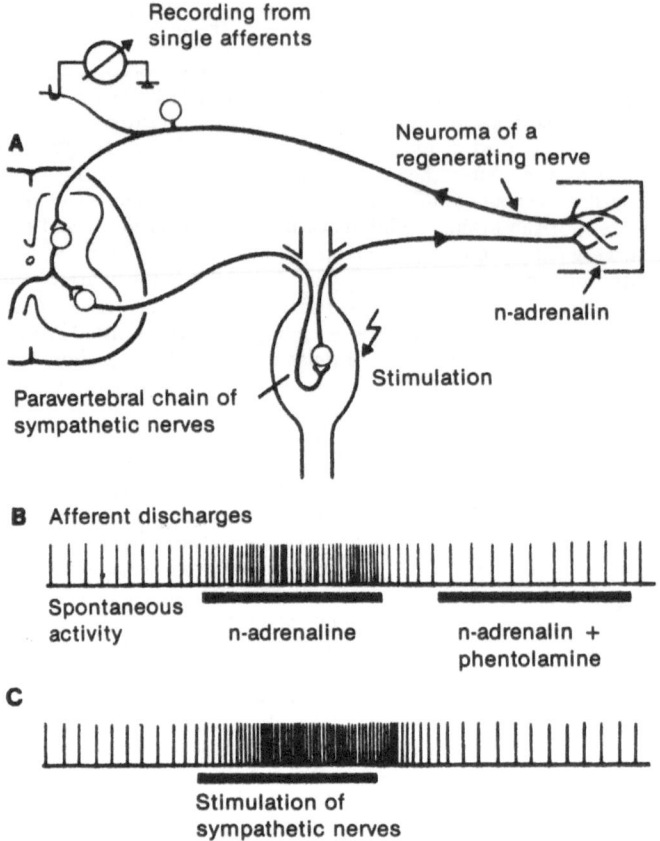

Fig. 2: Schematic drawing of action potentials of a regenerating rat nerve fiber (A). Spontaneous activity represents mainly A- and C- fibers (B; left). This is being increased by local or i.v. application of n-adrenaline (B; center), which effect may be reversed by alpha blocking agents (B, right). Direct electric "stimulation" of sympathetic fibers (C; center) also increases level of activity of afferent fibers over the spontaneous activity. From ZIMMERMANN [254].

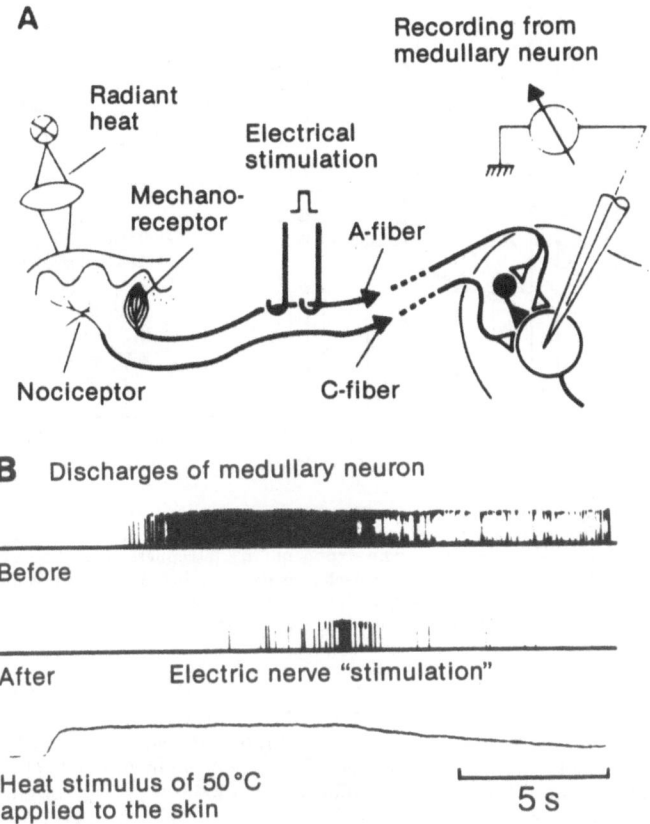

A

Recording from
medullary neuron

Radiant
heat

Electrical
stimulation

Mechano-
receptor

A-fiber

Nociceptor

C-fiber

B Discharges of medullary neuron

Before

After   Electric nerve "stimulation"

Heat stimulus of 50 °C
applied to the skin

5 s

Fig. 3: Inhibition of cat neurons by electric nerve block. ZIMMERMANN [254] uses the
term "electric nerve stimulation" in the original text. The author of this monograph however (JENKNER) sees the reduction of the number of action potentials
in the time unit following this "stimulation" as a most impressive demonstration of a true blocking effect. Therefore this TENS does not have the effect of
a stimulation but of a genuine nerve block.
(A) sketch of experimental situation: microelectrodes registering from single
medullary neurons. Radiant heat without and with simultaneous electric "stimulation" of A-delta fibers.
(B) Discharge of medullary neuron before (upper half) and after (lower half)
10 minutes of repetitive stimulation using 50 hz impulses. Lowest tracing is
equivalent of skin temperature. Time mark = 5 sec.

13

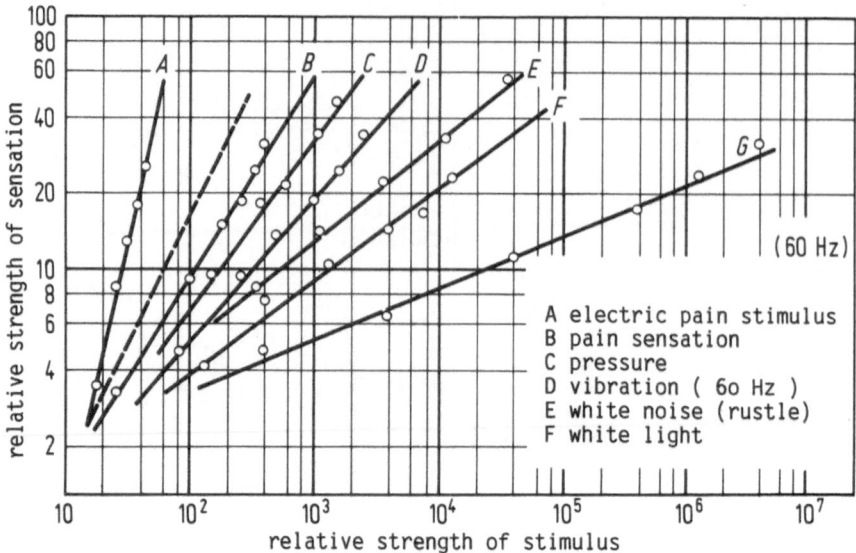

Fig. 4: Histogram on relative strength of sensation as a function of relative strength of stimulation for various sensory modalities, as seen from STEVENS' potential function $E = k \cdot R^n$. The exponent "n" (steepness of incline of potential function) is characteristic for each sensory modality. It is smallest for light and the greatest for pain [KEIDEL, 139; after STEVENS l.c. 139]. With kind permission of author and publisher.

Aside from these few but most important neurophysiologic pieces of information, one aspect of the field of neurochemistry seems important enough to point the attention of the reader to it for better understanding of what will be mentioned a little later. Recent findings demonstrated that patients suffering from pain of organic origin show a lowered level of one fraction of endorphins in the cerebrospinal fluid [2, 144] compared to normal (non- pain) subjects. This level will increase back to normal when pain is treated by brief electrical impulses. While patients suffering from pain of purely, or almost only, psychic origin will have a normal endorphin level in their CSF [2, 3]. After all, it is very well known that the psyche (soul, mind, brain) may initiate pain. It is very much less known that psychic factors also may inhibit the sensation of pain [JENKNER, 113]. The sometimes called miraculous behaviour of Hindu fakirs may be cited as a well known example, which had been demonstrated objectively by LARBIG [157]. Therefore patients with chronic pain from psychic causes demonstrate normal endorphin levels in their CSF and are to be assigned to the so-called "idiopathic pain syndrome" according to ALMAY [2]. This syndrome belongs to the depressive disorders; these patients should be treated accordingly.

One may surmise that the level of endorphins in the CSF as well as of other substances (such as e.g. substance P, the level of which also was seen changed in pain patients) may be of importance for the effect of this and many other forms of therapy. In lowered CSF levels, these therapies are very effective; normal levels seem to prevent any effect. This finding is most important, since empirical information for a long time indicated that this electric therapy as well as e.g. pharmacologic nerve blocks or neurosurgical interventions

14

lack any effect in psychogenic pain. For practical purposes this would mean that before starting any therapy patients with mainly, or only, psychic causes for their pain should be rejected and not admitted for this therapy. Of course, very unpleasant painful sensations of long standing influence the psyche of any patient; but these changes are reactive and not a causative factor. True relief of pain will decrease and finally remove these reactive psychic changes.

OPTIMAL FORM OF THERAPY

Sunt verba et voces, quibus hunc lenire dolorem
possis et magnam morbi deponere partem.
(Quintus Horatius Flaccus; Epistularum liber I,1,34)

As already mentioned, we expected different forms of electric impulses [4, 100, 213, 242, 243] to have differing effects on nerve fibers, if there is one at all, and not identical effects as studies of the literature indicated. On a model of sympathetic fibers, which are of small diameters, studies on a possible effect of an electric field established by skin electrodes were undertaken. Of course, the neurophysiologic differences between sympathetic and C-fibers had been noted, but in diameter and other criteria (see table 1) close relationships exist between the two types of fibers. Sympathetic fibers are very well suited, especially since they have an end organ which may be observed objectively: the vessels, which are not influenced by the method chosen for their evaluation, namely rheography [115, 136, 189]. If functional changes of the sympathetic system of fibers have to be made, one should be aware of the various effects the different states of this system have on vessels. Documentation on the changes in the end organs is presented in the following figures. In fig. 5 the result of direct mechanical stimulation of the sympathetic chain (approached operatively in a patient requiring section of this chain; who consented to physiologic observations during his operation) and of surgical section is presented. Fig. 6 shows the effect of sympathetic stimulation on cerebral circulation caused by placing ice cubes on the soles of the feet. Fig. 7 reveals the effect of a stellate block by a local anesthetic. With this information at hand, one may use skin electrodes to flood the stellate ganglion electrically. Fig. 8 and 9 show anodal and cathodal electrode positions respectively.

Fiber type	Function	FD um	NCT m/sec	DSI ms	aRP ms
A α	Proprioception somatic motor fibers	12-20	70-120	0,4-0,5	0.4-1,0
A ϑ	pain, temperature, touch	2-5	12-30	0,4-0,5	0,4-1,0
C-dorsal roots	pain, reflex response	0,4-1,2	0,5-2,0	2	2
Sympathetic	postgangl. sy.-fibers	0,3-1,3	0,7-2,3	2	2

Table 1: Fiber types in some vertebrate nerves with some data (partially reproduced from [63]).

FD = Fiber diameter   NCT = nerve conduction time
DSI = duration of single impulse   aRP = absol. refract. period.

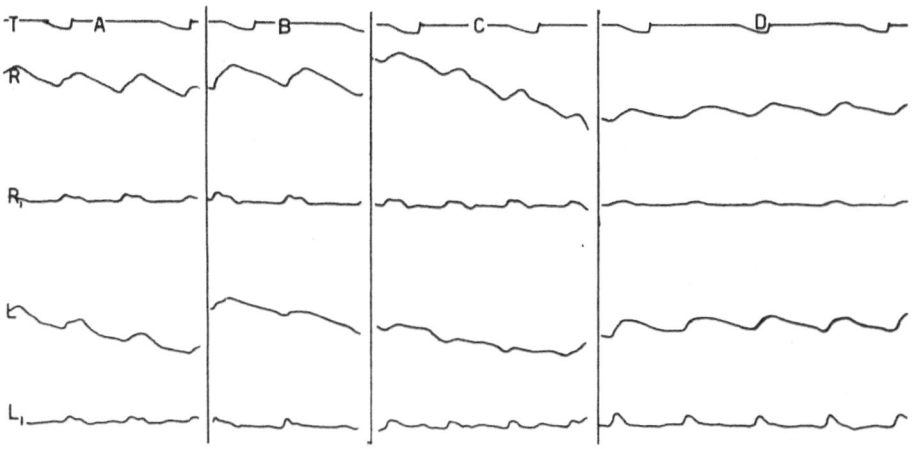

Fig. 5:   Effect of direct mechanic stimulation and surgical section of left cervico-thoracic sympathetic chain on rheoencephalogram (REG, which is an electric pulse wave recording of internal carotid artery tree; see JENKNER, [115].) (A) Sympathetic chain approached surgically with the patient sitting upright and fully anesthetized, starting condition before effectuating the section required for therapeutic reasons. (B) immediately after mechanic stimulation of chain on left. (C) 10 minutes after (B). (D) immediately following surgical sectioning of left chain. From JENKNER [110].

Symbols mean: T = time in sec, R(L) = REG of right (left) half of head with first derivative R_1 (L_1) of respective curve.

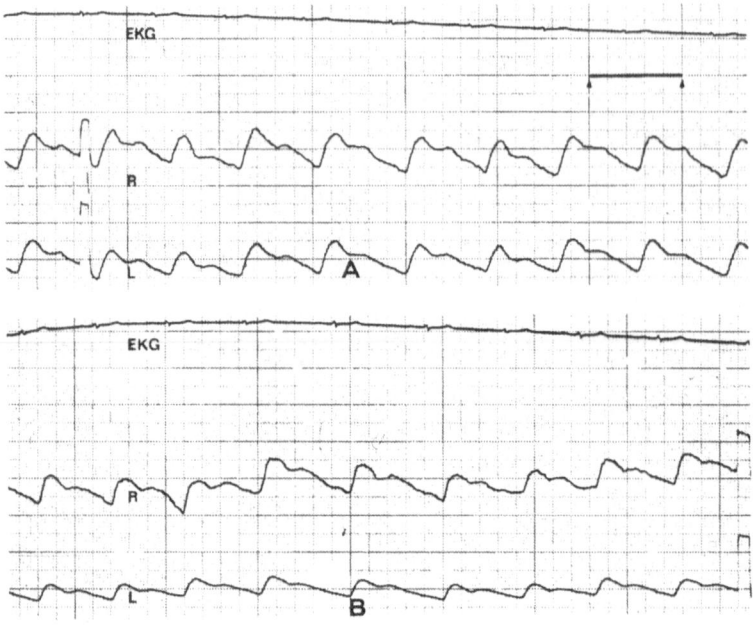

Fig. 6: Rheoencephalographic observation on cerebral hemodynamics before (upper curves) and after (lower tracings) a distant stimulation of the sympathetic nervous system, in this case by application of ice cubes onto the plantar faces of both feet). Symbols mean: EKG = electrocardiogram; R (L) = REG of right (left) hemisphere; time mark in seconds; amplitude signal = 100 mOhm. From JENKNER [115].
Amplitude of lower pair of curves much lower = spastic vessels due to sympathetic irritation.

18

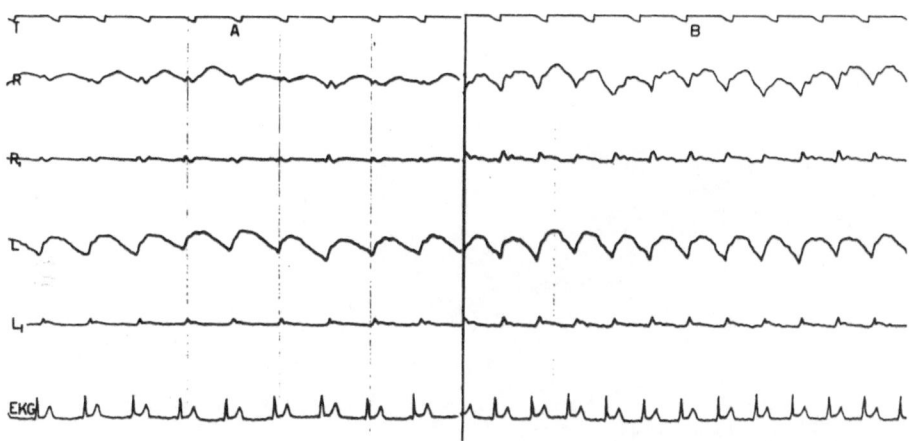

Fig. 7: Rheoencephaloggraphic tracing before (A) and after (B) a right stellate ganglion block using a local anesthetic (5 ml xylocaine 1% without adrenaline). From top to bottom: T = time in sec.; R (L) REG of right (left) hemisphere; $R_1(L_1)$ first derivative of R (L); EKG = electrocardiogram. Observe the relatively important increase in amplitude of curve R in (B) as compared to (A). From JENKNER [115].

Fig. 8: Anode in correct placement for blocking the right stellate ganglion. From JENKNER [118] (over leaf).

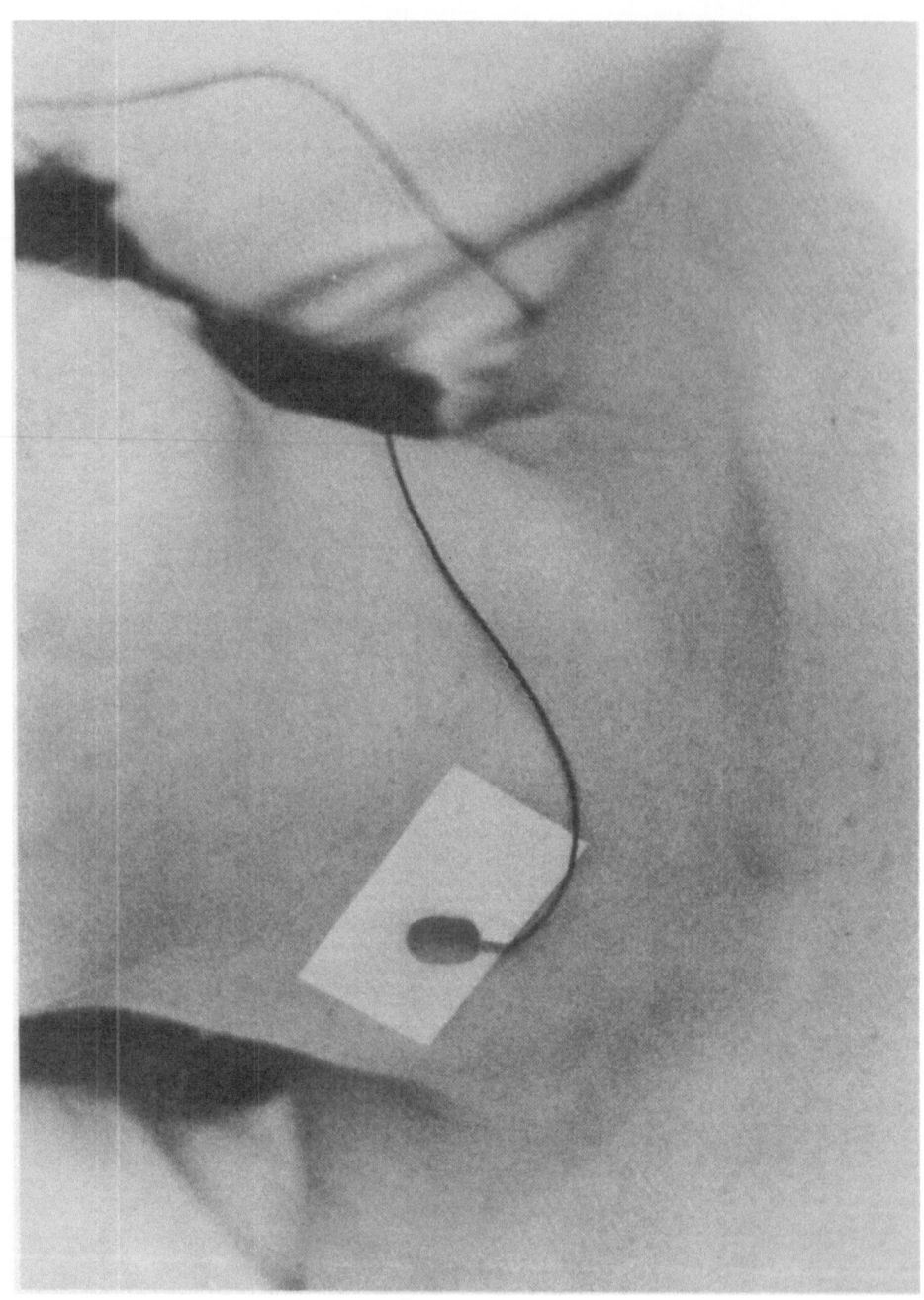

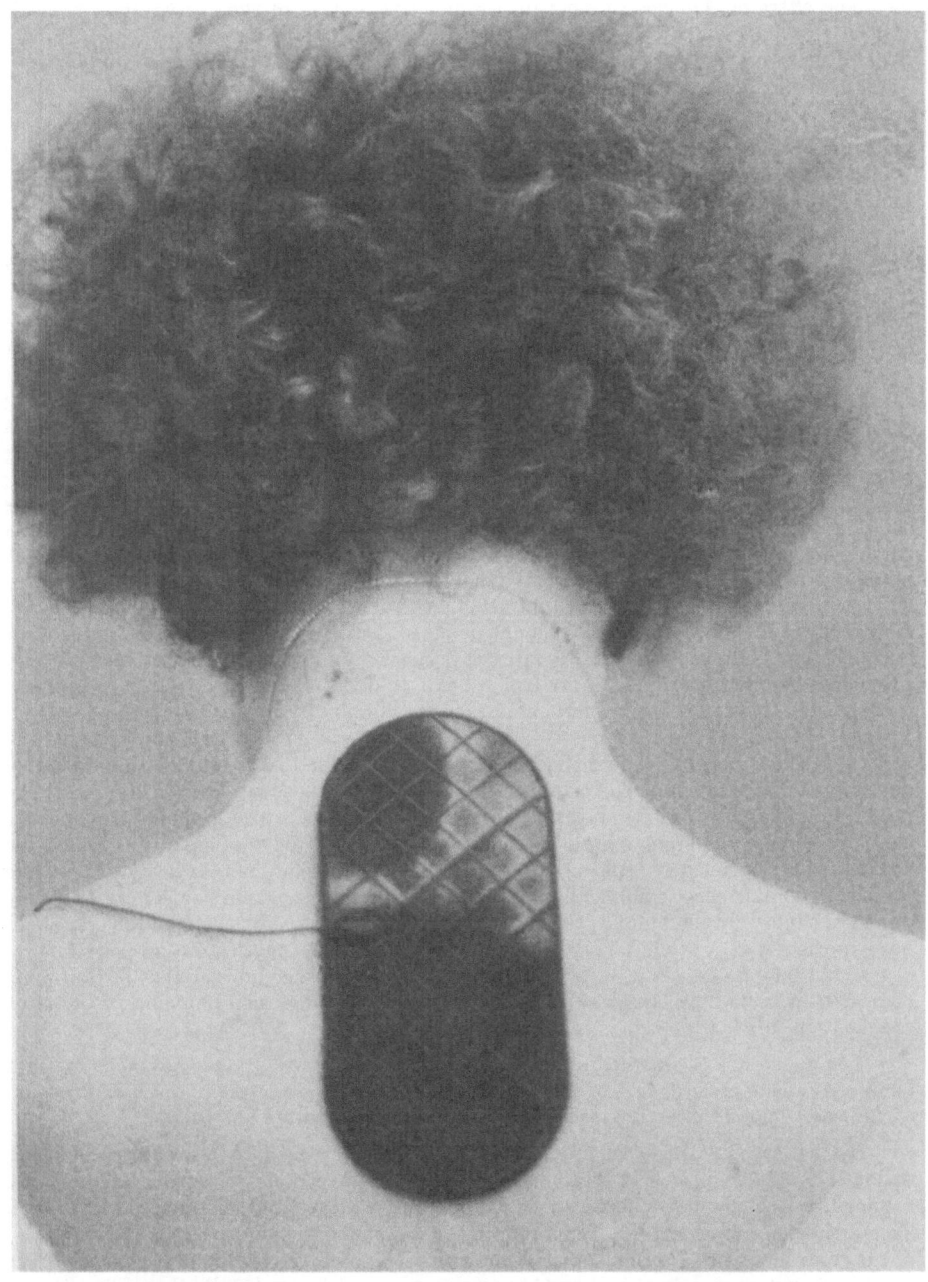

Fig. 9:   Cathode in place for either right or left sided (or bilate) stellate block. From JENKNER [118].

The effect of this electric field on the functional state of the sympathetic fibers is evident from fig. 10. The rheoencephalogram changes the pretreatment control record in exactly the same way as is known to occur when blocking the stellate ganglion by a local anesthetic (fig. 7).

This model, which was developed in animal experiments, which were then repeated on humans, allows us to compare various types of electric impulses, as well as sizes and positions of electrodes. In this way an optimal wave form and optimal electrode sizes and positions were determined. Electrode positions shown in fig. 8 and 9 are optimal for influencing the stellate ganglion.

Thus an optimal wave form was found [130]. It is characterized by the following criteria: DC-square waves of rather brief duration of a single impulse (under 0.2 ms, optimally 100 us) and peak current per pulse as high as possible (20 - 30 mA should definitely be reached); these impulses should be applied between 20 and 50 times per second (the mean being 35, which is quite practical, since human eyes have a flicker fusion frequency of about 35: visual regulation allows us to set this frequency by turning the frequency regulation to a point where one is able to still see the light flicker). This number of impulses per second was recognized already 1935 [27, 77, 213, 242] as being responsible for best effects on thin nerve fibers. After confirmation by this author (1977; [116] and 1980;[117] it was reconfirmed by JOHNSON (1991; [134]). The field has to be established from electrodes very unequal in size: field density is much greater under a small electrode (see fig. 11) which should be the anode. It should be placed as close as possible on the skin surface to the nerve one wants to block. Under an anode a soothing effect has been known a very long time [109] and hyperpolarisation ensues [KEIDEL, 138] disrupting conduction. A cathode should be placed on the surface of the body (trunk or extremity) immediately opposite the anodal position (with the exception of occipital and trigeminal neuralgias and some rare conditions). It should be relatively large to decrease field density under it to such a low value that no stimulatory influence may be detected. Duration of daily therapy should be 20 minutes; a longer interval does not increase the effect and the extent of optimal effect is reached only after 15 minutes. Optimal time of day is just before maximum of pain sensation would be expected. A check-up with the prescribing physician after every 5th therapy session or once a week in home treatment is desirable. Using the bipolar form of impulse (AC) or electrodes of equal size should be avoided. The reason why biphasic forms of impulses are of no help is easily given: The positive, desirable effect coming about during an anodal phase would be destroyed entirely during the interval this same electrode is serving as a cathode. At the end of the treatment interval the net gain would be zero. To allow the prescribing physician a proper selection of optimal impulses a selection of apparatus available commercially in central Europe is presented in table 2; all technical details necessary for proper identification of impulses are mentioned.

The purpose of this monograph with its orientation toward practical aspects does not allow a complete presentation of all present theories of pain perception nor an explanation of the exact manner of action of these stimuli, even if they were known.

By the reason of the exact proof that these selected types of impulses have a blocking action it should be proposed that the term "TENS" (for Transcutaneous Electric Nerve Stimulation) should be replaced by " TENB" (for Transdermal Electric Nerve Block). This is the place to repeat that only impulses of a monophasic nature should be employed in the control of pain. No single author has been able to objectively demonstrate an effect of biphasic pulses on pain reduction (MANNHEIMER and LAMPE, [174]). Newer Studies try to find a certain type of impulse as beeing optimally suited. JOHNSON et al, [132] could

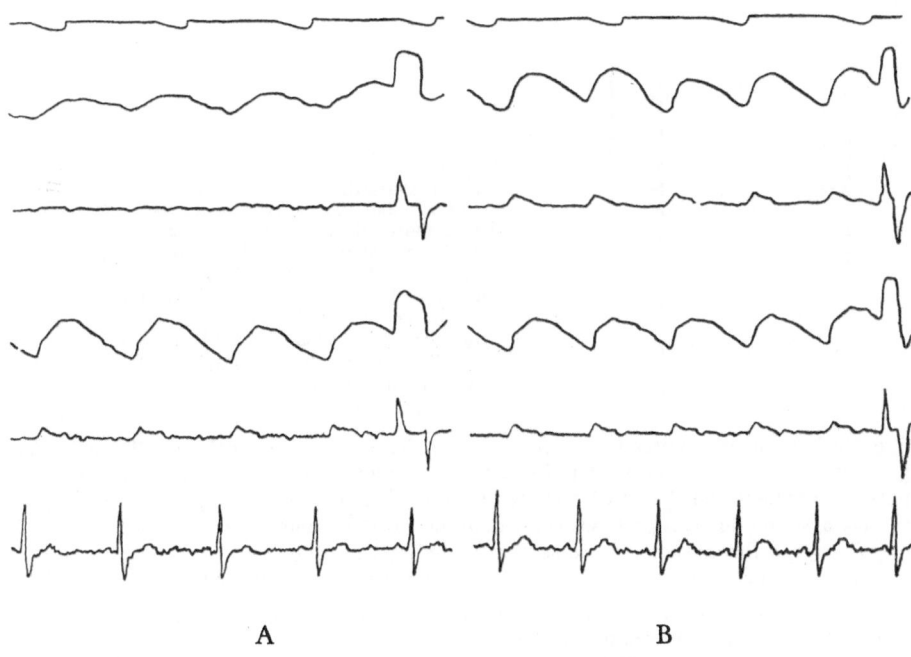

A   B

Fig. 10:   Rheoencephalograms (A) before (B) after a right electric stellate block of 20
minutes duration. From JENKNER [116].
Sequence of curves as in fig. 5, exept: bottom tracing is electrocardiogram.
Explanation see text on pg 21, first paragraph.

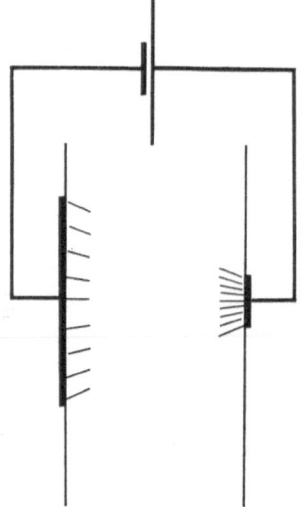

Fig. 11:
Schematic drawing of density of field lines under a small and a large electrode under assumed identical current characteristics. Note that under small electrode, field lines are considerably more dense (closer together) and a nerve being positioned under this anode will be influenced by the field much more than if placed under the large electrode (where field lines are farther apart). From JENKNER [118].

find – using biphasic pulses with a frequency of 70 and a pulse width of 200 μsec – that patients liked a certain type best, for reasons of comfort; but these were different from patient to patient and stayed without any connection to neurophysiologic facts of the pain sensing system. But it is just this connection between neurophysiology and the sensation of pain, which this author tries to stress: If one is trying to treat pain one should adhere to the laws of neurophysiology unless one is doomed not to obtain optimal results.

These laws request and certain recent findings indicate [237] that these impulses need to be modulated in one form or another. This postulate was actually met by the authors' favorite impulse ever since 1977 [114].

It is certain that smaller electrodes act better than larger ones in reducing pain (JOHNSON et al, [134]). NOLAN [197] examined the conductivity of 25 different types of electrodes. Beause their impedance is varying widely (between 1000 and 7000 ohms) he concludes that this must have effects on basic research and clinical effects. JENKNER uses electrodes of an impedance of less than 200 ohms (silicone with different types of caraya {natural = 87; T-20 polymere = 57; Lectron II = 40 and Spectra 360 gel = 37}); the new type of adhesive electrodes (PALS) were mesured to have 12 – 280 ohms impedance; these studies were carried out with the optimal type of impulses, not using DC and avoiding total tissue impedance; the latter largely depends on electrode distance and is mentioned on pg 30. Thereby optimal current flow is guaranteed.

Only JENKNER and SCHUHFRIED [130] have compared objectively monophasic and biphasic impulses using the rheogram. All biphasic pulses, also the asymmetrical ones, showed no effect on the rheogram if tested after the end of current flow and the sympathetic fibers had been subjected to the field (JENKNER, unpublished studies). This author also had compared various instruments supplying different types of impulses on their effect to reduce pain (unpublished). It was found that biphasic sets never reduced pain for more than 30 – 50% during current flow and that this effect vaned soon; three months after

24

therapy pain reduction had diminished to a mere 5 - 10%. JOHNSON et al, [132] reported on 170 cases using biphasic pulses that the effect manifested itself already 1/2 hour after start of current flow (in 75% of patients) but disappeared just as fast already 1/2 hour after termination of current flow (51% of patients). Only 47% of patients reported 50% reduction of pain only during current flow.

apparatus used (coded)	studied on ...cases	pain reduction in %	
		at end of therapy	3 months after end of therapy
biphasic-symm. or biph.-assym. apparatus (all forms and frequencies)			
ES	45	35	5
N5	38	53	7
ME	19	54	10
BP	11	65	8
purely monphasic apparatus (35 impulses/sec and square wave only)			
RX	927	70	65
EA	316	70	62
NG	85	72	60
SP	115	72	67
R2	76	74	70
E5	43	88	84

(all apparatus represented in table 2; code not disclosed)

Statistical evaluation of these data showed that the pain reduction using biphasic curents was statistically significantly lower after 3 months than immediately after end of therapy (t-test:p= 0.000391), while such a difference for monophasic currents lacked (p= 0.0934) statistical significance. While the difference between pain reduction immediately after therapy was significantly different from biphasic to monophasic results (t-test: p=00187) the significance was more expressed after three months (p=0.0024). Calculations were performed by BLAISE statistic program and comprised Fisher exact test and t-test.

On the other hand, with exact monophasic impulses [126] of named frequency, electrode size and electrode positions (see specific section of monograph) reducing pain if caused by benign processes reached more than 80% and in a malign causation 60% and these results are obtainable by all physicians observing the named specifications of impulses. Therefore, we can not advocate the use of other than monophasic impulses for the reduction of pain. Nevertheless, also apparatus delivering other impulses are mentioned in table 2 mainly because they are being offered frequently. The interested reader shall now be informed in brief on the most interesting theories of pain after a brief regard for the question of energy.

Table 2:

Manufacturer or representation*	apparatus designation (type)**	form	m sec duration	cps frequency***	mA current
			of single impulse		
Agar Electronics and	Neurogar II(A)	bi-ph.ass.	0.25-0.5	4-120	45-70
Metal Products Inc.,	Neurogar III(A)	bi-ph.ass.	0.25-0.5	4-120	50
Ginosar,14980 Israel	Neurogar IV(B)	bi-ph.	0.25-0.37	80	-50
	Neurogar TNS(C)	bi-ph.ass.	0.25-0.37	80	-20
(Science Trading HG, Glauburgstraße 66, D-60318 Frankfurt/M., FRG)					
American Medical	Nurite I(B)	mono	0.7-4.0	30-120	-55
Systems Inc, 4001	Nurite II(B)	mono	0.7-4.0	30-120	-55
Stinson Blv.NE.,Minneapolis, MN., 55421, USA					
(Agency: FRG: Kontron GmbH, Oskar-von- Millertstraße 1, D-85386 Eching)					
AVERY, 145 Rome Street	TNS-T10(A)	bi-ph.	1.0-4.0	20-200	0.5-3.0
Farmingdale, NY 11753, USA,					
(Institut Straumann, CH-4437 Waldenburg, Switzerland,					
Avery GMBH, Kollanstraße 105, Hamburg 61, FRG)					
Berryfast GmbH	BerryMed(A)	bi-ph.ass.	0.2	2-120	-80(cave:A)
Postfach 1109	BM 9	(modulat)	0.06-0.2	10-100	-60
D-32602 Vlotho, FRG					
BIOMETER, Thorsgade 8	ELPHA 500(C)	mono	0.2	1-4	-60
DK-5000 Odense C, Denmark	TENSMED II(C)	mono	0.1-0.3	1-100	
(A: MTS Medizintechnik Skola,	ELPHA 2000(C)	mono M	0.1-0.25		
A-1210, Áugelgasse 6)	10-100	-31			
		mono II	0.07-0.2	10-150	-38
		mono I	0.05-0.25	10-150	-31
		mono burst	0.15-0.25	0.05-5	-18
CEFAR, Mantalskroken 8	CEFAR-Dumo(B)	mono	0.17	conv.50-120	-60
S-222 47 Lund Sweden				burst:1.7	
(FRG:Tettamed GmbH,	CEFAR-Dual(B)	mono	0.2	conv.10-100	-60
Dieselstraße 14, D-50859 Köln 40)				burst:1.7	
T.H.Charters Inc.	ElectroBLOC	mono	0.1	35 b	-30
9985 S.W. Heather Lane	(C)	(surge)			
Beaverton, OR. 97005		Mono FM	0.2	3/30	-40
USA	NeuroPulser 2+(C)	bi-ph.ass.	0.03-0.25	0.8-100	-30
(A: MTS Medizintechnik Skola,	NeuroPulser 2+(C)	mono	0.03- 0.25	0.8-100	-30
A-1210, Áugelgasse 6)	Mini-1(C)	bi-ph.ass.	0.24	70 b1	-20
	Mono-BLOC(C)	mono S	0.1	35 b	- 28
		mono FM	0.2	3/30	-32
		bi.ph. MsR	0.3	0.66	-28
Christian Nissen	TNS 821(B)	mono	0.2	2/100	-50
Medical Pharmaceutical Co.Ltd, PF 325, SF-00101 Helsinki 10					
CODMAN and	EPC(B)	mono	0.5-3.0	15-180	-50
SHURTLEFF Inc.,	EPC dual(B)		0.5-3.0	15-180	-50
Randolph,	EPC mini(B)		0.5-3.0	10-100	-60
MA 02368, USA	EPC MINI dual(B)	mono	0.5-2.5	10-90	-60
	STIMETTE 6065(B)	mono	0.5-2.5	3-150	-45
	STIMETTE (A)	bi-ph.ass.	0.5-2.5	3-150	-45
	STIMETTE 6066	mono	0.5-2.5	3-150	-50
	STIMETTE 6067	mono	0.5-2.5	3-150	-50
Dan-Sjö	Neurostal	bi-ph.ass.	0.2	40-120	
Electronik AG Box 144 Sundyberg, Sweden					
DOLTRON A.G.,	ESA 400(A)	bi-ph.	0.1-12%(I)	4-120	5-55
Webergasse 5	ESA 600(A)	bi-ph.	0.1-12%(I)	4-120	5-55
CH-8610 Uster,	ESA 1000(A)	bi-ph.	0.1-12%(I)	4-120	5-55
Switzerland	TNS 200(A)	bi-ph.	0.1-12%(I)	4-120	5-55
3M Inc. St.Paul,MN	TENZCARE 6260(B)	mono	0.02-0.3	10-200	-75
55101, USA.				b: 20	-45
(FRG: 3M Medica	TENZCARE 6215(B)	bi-ph.ass.	0.024-0.25	c: 2-170	-80
Wilbecke 12-14,				b: 100	
D-46325 Borken	COMPTENZ(A)	bi-ph.ass	0.025-0.25	c: 2-140	-60
				b: 2	
	OPTENZ(A)	bi-ph.ass.	0.1	100	-50
Electromedical	Alphastim 2000A	bi-ph.ass.	0.5	25-320	-50
Products Inc.,	Alphastim 2000B	bi-ph.ass.	0.075-2.0	3-80	-50
Hawthorne, CA	Alphastim 3500	bi-ph.ass.		0.5-120	-50
Electronic Research	Micropulsar(B)	bi-ph.ass.	0.025-0.25	3-140	-60
Devices Corporation;	Model 8803				
9320 S.W. Barbur Blvd, Portland, Oregon 97219, USA					
ENRAF NONIUS, Inc.,	EM-SET(A)	mono	0.2	100; 2	-50
Delft, Netherlands	TENSMED II(C)	mono	0.05-0.25	2-100	-28
	TENSMED I(C)	mono	0.05-0.2	2-100	-28
GENERAL MEDICAL	Neurological	mono	?	3; 30	
INDUSTRIES Inc.,	Stimulator				
969 Barcelona Drive, Santa Barbara, CA 93106, USA					
ITO Co.Ldt., Japan	Energy 30(C)	mono	0.01	30	H 20
FUTUREMED Inc., USA					L 10
	Energy 4(C)	mono	0.01	4	H 20
A: Dr. F. Schuhfried, A-1090					L 10
Van Swietengasse 10	DR.PULSE(C)	mono	0.1	1-60	16

open circuit voltage	channels	electrode sizes (i,d)****	source of energy (B/rech)	size cm/cm/cm	approx price (US $)*****
-45	2	i	+(4x1.2)+	12.0x8.3x3.9	550,-
-50	2	i	+(4x1.2)+	12.0x8.3x3.9	950,-
-50	2	i	+(4x1.2)+	12.0x8.3x3.9	1.500,-
	1	i	+		800,-
	2	i	+		950,-
-100	1/2	d	+		520,-
-110	2	i	1x9V	9.1x6.4x2.3	?
-145	1	d	(9V)+	11.0x6.6x3.4	500,-
	2	d	9V	11.0x6.6x3.4	280,-
	2	i	+/+	11.5x6.0x3.0	
(Timer 20-100 min; all Data on LED-Display 20x36 mm)					
140	2	i	+	10.0x6.5x2.5	?
140	2	i	+		?
-100	2	d	+ (4x1.5V)	15.0x17.0x6.3	895,-
(Timer 10/20 min)					
-120	2	d	+/+	8.2x6.4x2.9	495,-
100	2	d	+(2x1.5)+	8.2x6.4x2.9	535,-
80	1	d	+/+	5.8x3.8x1.8	395,-
	1	d	+/+	11.4x6.8x2.6	595,-
(automat. timer 20 min)					
	1	i	2x9V	11.5x6.4x2.7	330,-
	1		+	7.0x4.0x1.7	not in Europe
	2		+	7.0x4.0x1.7	not in Europe
	1		+	7.0x4.0x1.7	not in Europe
	2		+(3x1.5)		700,-
	1		+(3x1.5)		700,-
	1		+(3x1.5)		700,-
	1		+(3x1.5)+	10.5x6.8x2.3	750,-
	2		+(3x1.5)+	10.5x6.8x2.3	750,-
-240		i	9V+	11.5x6.5x3.0	?
+/-47	4	i	mains		4.750,-
+/-47	6	i	mains		5.500,-
+/-47	10	i	mains		6.000,-
	2	i	mains		?
	2	d	(4x1.5)+		
	2	d	(2x1.5)+	9.0x5.9x2.3	
	2	i	(2x1.5)+	9.5x5.9x2.4	
	2	i	(2x1.5)+	6.35x5.8x1.68	
14	2		9V		?
	2		12V+		?
19	2		12V+		?
	1	i	1x9V+	7.9x5.7x2.3	?
?	1	d	+	11.0x7.5x2.4	400,-
84	2	d	+/+	11.0x7.5x2.4	400,-
-84	1	d	1x9V	11.0x7.5x2.4	270,-
60	1	i	+/+		?
100	1	i	3V	φ 2.5x0.9	60.-
50					
100	1	i	3V	φ 2.5x0.9	60,-
50					
	1	i	9V	11.3x5.7x2.2	not in Europe

Table 2 (continued)

Manufacturer or representation*	apparatus designation (type)**	form	m sec duration	cps frequency***	mA current
		of	single	impulse	
KOVOPODNIK Senovazna 4 CS-111 98 Prag 1, CSFR	Analogic(C) Analogic Klinic(B)	mono mono	0.3-3.1 ?	3-150 ?	0-30 ?
Krieger Gesundheits- technik, Schubertgasse 22 A-1090 Wien, Austria	TNB-K3(B) TNB-K4(C) TNB-K5 Automat(C)	mono mono mono	0.05 0.2 0.2	2-90 1-100 1-100	40 30 23
LA JOLLA TECHNOLOGY Inc., 11558 Sorrento Vally Road, San Diego, CA 92121, USA	Dynex II 2005(A)	mono + net DC	0.04-0.2 0.4-2.0	2-110 14	-60
MEDTRONIC Inc. 3055 Old Highway Eight, POB 1453 Minneapolis MN 55440, USA	Selectra 7750(A) Selectra 7720(A) Neuromed 3723 Eclipse 7723(A) Comfort Wave 7721(A)	bi-ph.ass. bi-ph.ass. bi-ph.ass. bi-ph.ass.	0.05-0.25 0.05-0.25 0.03-0.25 0.05-0.25	2- 99 2-99 2-125 85	-60 -60 3-65 -60
(Agencies: Europe: Neuro-Dpt. 60 Medtronic Sa. 120 Avenue Charles de Gaulle, F-92200 Neuilly sur Seine, Paris, France FRG: Medtronic GmbH, Kieler Straße 212, D-22525 Hamburg . A: No representation. Contact: Krauth & Timmermann, Poppenbutteler Bogen 11, D-22399 Hamburg, FRG.)					
EUREX Inc. POB 2805, Portland, Oregon, 97208, USA	Bipulse(A)	bi-ph.	0.075	10-100	-95
NEEN PAIN SYSTEMS Old Pharmacy Yard Church Street, East Dereham, Norfolk NR 19 1DJ, GB	Micro TENS(B) XENOS 200(B)	bi-ph.ass. bi-ph.ass.	0.2 0.09-0.3	15-175 b 1-150	-50 -50
Neuromedics Inc. 10518 Kingshurst Houston, Texas, USA (FRG: Science Trading HG, Glauburgstraße 66 D-60318 Frankfurt/M. FRG)	Microceptor I(B) Microceptor II(B) Miniceptor I(B) Miniceptor II(B) Maxiceptor(B) Microceptor(A) Ultrastim 650-01(A)	bi-ph. bi-ph. mono mono bi-ph. bi-ph.ass. several: mono bi, triphasic mod.	1.2-2.0 1.2-2.0 0.4-1.0 0.41.0 0.5-4.0 0.2 0.1-5.0 1.5-2.0	10-100 10-100 25-100 25-100 10-100 120, 14 1-70 b (1100-2000)	80-100 -85 -100 -90 129 -150 -99
NORDAMED Wetzelstraße 8, Hamburg 60, FRG	Minimedical(A)	mono	0.8	4-60	50
Pain Suppression Lab. 1200 Route 46, Clifton, New Jersey, USA, 07 407	Pain suppressor(?) GL 105 C	?	8-20 12-20kHz	20	
REDA AG CH-6045 Meggern, Switzerland	MM 1000(C) MM 2000(C)	mono mono	0.2(0.1-0.3) 0.2(0.1-0.3)	2-120 2-120	33 33
A: Bständig KG., Lerchenfelder Straße 88, A-1080 Wien, Austria					
Dr. Schuhfried Van Swietengasse 10 A-1090 Wien, Austria	Stimulette(C) Relaxette(C) Dolorette(C)	mono mono mono	0.1 0.1 0.1	35 b. 35 b a. 3 35(b), 200	-110 -70 120
schwa-medico GmbH Büro Lahn, Wilhelmstraße 19	TNS SM1(C) TNS SM2(C) (also other sets: SM-1B, SM 2 MF, SM 2L, SM 1F)	mono mono	0.2 0.2	2-100 2-100	24 24
D-35398 Giessen, FRG [also in Frankfurt, Hamburg, Düsseldorf, Karlsruhe, Saarbrücken; Braunau]					
SHARP KK Nagaike-cho 22-22, OSAKA-SHI, Abeno-ku, Japan	JOYUP MH-100(C)	bi-ph.ass.	0.15	2-56	3.3
Shinsei Syoji Ltd. Tokio, Japan	Tiger Pulse(?)	mono	0.2	1.5-27	5.5
SPEMBLEY LTD. Newbury Road, Andover, Hampshire SP 10 4DR, England	9300(B) 9000(B)	bi-ph.ass. bi-ph.ass.	0.25 0.2	80 15-200, 2	-50 -50
(FRG: Science Trading HG, Glauburgstraße 66, D-60318 Frankfurt/M., FRG)					
Top Surgical MFG. Co. Ldt.; 4-6 Senju Tatuta - cho, Adachi - Ku, Tokio, Japan	POLE NEUTRACER position: high position: low	mono	0.1 1.4	1.5-50 1.5-40	
Agencies: Europe: LABAZ, Equipe Medical, B.v., Postbus 97, 3140 Maasluis, NL FRG: Medizintechnik Dr. Lotz, Säbenerstraße 70, D-81545 München, FRG A: Chemomedica, Wipplingerstraße 19, A-1013 Wien, Austria (needle only).					
Vana GmbH., W. Schmälzlgasse 6 A-1020 Wien, Austria	VANA I(C) VANA IV	mono mono+bi-ph.	0.05-0.3 0.1-0.8	1-200 6-400	5-40 5-40

*   Companies in alphabetical Order.
** Values of load tested were: (A) = 500; (B) = 1000 and (C) = 3000 ohm.
    Voltages in cases of biphasic shape are peak-to-peak.
    Attention: testing with less than 3000 ohm load gives unrealistic value
    of mA and V for clinical use.
*** b = burst; c = conventional

open circuit voltage	channels	electrode sizes (i,d)****	source of energy (B/rech)	size cm/cm/cm	approx price (US $)*****
-40	1	d	+(4.5V)	13.0x8.2x3.3	120,–
?	3	d	mains		?
	1	d	+/+	14.5x8.3x3.3	
	1	d	9V	14.5x8.3x3.3	400,–
	1	d	9V	14.5x8.3x3.3	?
-40	2	i	batt.pack 5.6V	6.0x8.3x1.9	600,–
	2	d	+4x1.5		615,–
	2	d	+4x1.5		605.–
-125	2	i	+1x9V	9.4x6.1x2.8	400,–
	2	i	+4x1.5+(AAA)	9.4x6.1x2.8	
	1	i	Akku 2,5+	5.7x6.3x1.6	?
	1	i	+/+	6.0x6.0x2.3	?
	2	i	+/+	7.3x5.4x2.7	?
-100	1	d	+/+		700,–
-85	2	d	+/+		900.–
-100	1	d	+/+		830,–
-100	2	d	+/+		830,–
120	2	d	+/+		850,–
	2	i	power supply unit		?
200	2	i	1x9V	14.8x9.2x3.6	?
15	1	i	9V		570,–
	1	i	B.		?
-62	1	d	9V	11.0x7.5x2.4	300,–
-62	2	d	9V	11.0x7.5x2.4	460,–
-350	1-3	d	mains		2100,–
-360	1	d	+(5 mono)		450,–
700	1-4	d	mains		4.420,–
	1	d	9V	3.5x6.5x11.0	500,–
	2	d	9V	3.5x6.5x11.0	630,–
110	1	i	+(4x1.5)		not in Europe
-50	1	i	+(9V)+		220,–
-20	2	i	+/+		840,–
-25	1	i	+/+		?
60	1	d	+	without acc.	480,–
15	needles				
60	1	d	+(9V)+	11.0x6.6x3.3	350,–
6-50	4	d	mains		3.300,–

**** i = identical size; d= differing sizes of electrodes available
***** figures without warranty.

It is postulated that the sale of these and similar apparatus be limited to "on prescription only".
Respective measures of government agencies should be undertaken to ascertain the safety of patients and
prevent financial loss such as may occur in case of unnecessary purchase of ineffective apparatus
e. g. in supermarkets, mail order business or from travelling salesmen.

THE QUESTION OF ENERGY

Every physician prescribing TENS units should have exact information on the technical details of the devices. Data on current (at a load of 3 kOhm, not less!) have to be supplemented by duration of a single impulse (200 μsec or less). One should use devices only if they provide electrodes of very unequal sizes. See section K of appendix. The shape of impulses should be monophasic square waves; biphasic impulses or asymmetrically biphasic impulses should not be prescribed. Every unit should be provided with a scheme of sensory distribution for allowing marking of electrode positions. Use red pencil for anodal and blue color for cathodal positions. Also mark these positions on the skin of patients! The skin mark should be renewed if the color disappears. Examples of sensory schemes are given under section J in the appendix.

After placement and fixation of the electrodes, the intensity of the current needs to be increased gradually. The patient will have to indicate when the typical sensation of flowing current is starting and when it is increasing to a level not yet uncomfortable. Patients being overcautious need to be told that increasing the intensity to an uncomfortable level is required only to decrease intensity until it is not uncomfortable. After 5 - 8 minutes of therapy, the subjective sensation of the intensity of the current will have decreased. Intensity needs to be increased again and again from time to time during the entire treatment period of 20 minutes. The goal is to have a maximal intensity of the current still comfortable for the patient to ensure largest possible field strength under the anode. Results depend on this energy level. Actually, the charge of electricity (analogous to the volume of water) is responsible for the physiologic effect. It is measured in Coulomb. A current of 1 amp flowing 1 sec moves the charge of 1 Coulomb (= 1 ampère-second). In the practical instance of TENB, assuming a current of 30 mA would flow with a duration of an impulse of 200 μsec, a charge of 6 micro- coulomb would be transported. For reasons of safety of the patient, the value of 25 μC should not be surpassed. A comparison between standard TENS and TENB shows (see table 2a) that the charge achieved by TENB is higher than with traditional TENS. Both are minimal compared to other modalities of therapy or medical applications [125].

type of electrotherapy	microcoulomb/pulse
external defibrillation	16.000 - 200.000
external cardiac pacemaker	200 - 2.000
electroconvulsive therapy	900
electric nerve block (TENB)	2 - 25
traditional TENS	2 - 15

Table 2a: Examples of typical charge per pulse, comparing various forms of electro-medical applications. From CHARTERS [42].

At this time, the assistance of the patient in determining the suitable intensity is definitely required. However, our studies have resulted in an array not requiring this help any longer. A gradual increase in intensity of the current is interrupted at the moment when a physiologic characteristic indicates that unpleasant sensations of the patient would soon be reached. This fully automated circuit will also take over the increase of intensity required from time to time which so-far needs manual adjustment by the patient or supervisor. Such an automated circuit ist not available commercially at present. The time at which it may become available is unforseeable. This development would be applicable to all electro-medical instrumentation and not only TENB. At other applications it would prevent untoward sensations, burning of skin and other tissue damage.

THEORIES OF PAIN

Phantasy looks at pain by a convex,
stoicism by a concave lens (Jean Paul)

The **Gate Control Theory** (MELZACK AND WALL, 1965, [182]) postulates a pain transmission gate located in the substantia gelatinosa region of the dorsal horn of the spinal cord. The theory states that pain transmission through this gate is altered by sensory nerve input to this region. Subsequent research since 1965 has supported much of the Gate Control Theory and has added new dimensions to it (WOLF, [250]; 1984; FIELDS AND LEVINE [66]; 1984). TENS pulse frequencies between 75 Hz and 150 Hz (cycles per second or pulses per second) are recommended to activate this mechanism (MANNHEIMER AND LAMPE, [174]; 1984). This theory would explain pain reduction only during flow of current.

Central Biasing Theory. Melzack and Fox proposed an expansion of the Gate Control Theory to emphasize a higher biasing mechanism in the brainstem and brain (MELZAK, [182]; 1975; FOX AND MELZAK, [70]; 1976; MELZACK at al., [183]; 1983). The original Gate Control Theory did not adequately explain pain relief produced by high intensity stimulation. Brief, intense stimulation was proposed to activate the inhibitory central biasing system within the brain stem.

In addition, chronic pain was thought to result from self-exciting neuron chains or recurrent neuron loops, setting up memory-like processes maintaining pathologic pain. High intensity stimulation may disrupt and reorganize the recurrent neural chains and thereby effect a more prolonged pain relief of weeks or even months. One author reports pain relief lasting more than two years (JENKNER, 1983, 1986, [118, 120].

Endogenous Opiate Theories. Endogenous opiates have been discovered in the CSF which act in a manner similar to morphine in the control of pain (PERT AND SNYDER, [200]; 1973; HUGES et al. [102], 1975). These substances, primarily the endorphin and enkephalin families, bond to receptor sites in the central nervous system and act as neurotransmitters and neuromodulators. The endogenous opiate systems are also proposed to modulate pain by activating descending inhibiting pathways which involve nonopioid (serotonin) systems. Electrical stimulation via TENS devices has been found to increase the quantities of endogenous opiates found in the central nervous system (SJOLUND et al., [224], 1977. Stimulation frequencies lower than 10 Hz have been found effective in elevating endorphin and enkephalin levels (MANNHEIMER AND LAMPE, [174], 1984).

Other authors LÜBEN et al. [171] could show, on the other hand, that endorphins are of no avail for explaining the effect of electrostimulation.

Peripheral Nerve Block. In this theory, TENS is functionally analogous to conventional local anesthetics which attenuate transmission of pain afferents. Electrical current has been found to block nerve conduction (GREGOR AND ZIMMERMANN, [79], 1973; ZIMMERMANN, [252], 1972). Electrical stimulation may produce an antidromic blockade slowing conduction velocity in the affected underlying neurons (CAMPELL AND TAUB, [38], 1973; IGNELZI AND NYQUIST, 1976, 1979, [105], [106]). V. KRAUTHAMMER [149] has done some basic studies on leeches. His conclusion is: " . . . if an external field induces a conduction block, it does so by membrane hyperpolarization at the branch point."

Sympathetic Nerve Block. This theory states that the sympathetic nervous system responds to certain types of electric stimulation with the result that sympathetically maintained pain is reduced (JENKNER AND SCHUHFRIED [130], 1981; JENKNER [118, 125], 1983, 1989).

The proper stimulation waveforms, pulse widths, frequencies, polarities, electrode locations and diagnosis are said to be vital to the success of this method. JENKNER, 1989 reports, that in most cases long-lasting pain relief is achieved with treatments of only 20 minutes per day for two weeks [125].

According to JENKNER AND SCHUHFRIED [130], 1981, the optimal stimulation waveform for pain control consists of bursts of monophasic pulses having waxing and waning amplitudes. The ElectroBLOC produces this optimal waveform (see Table 2 under: Charters).

S u m m a r y . Health and Human Services Publication FDA 86-4209 summarizes the status of TENS theories as follows: "Indeed, the experimental evidence is contradictory and not yet compelling for any of the . . . theories. The most consistent observation between researchers and across publications is the nearly unanimous statement that definitive information is still lacking on the precise neurophysiological mechanisms underlying TENS pain relief (SORIC AND DEVLIN, 1985). . . . Thus, each theoretical mechanism is neither individually sufficient nor necessary to fully explain TENS-induced pain relief. At best, each potential mechanism may be conditionally sufficient but not universally necessary to account for TENS analgesia, dependent upon patient, pain and stimulation parameters. We should hasten to add, however, that the lack of satisfactory theory does not disprove either the reported safety or relative effectiveness of TENS analgesia. Rather, the literature suggests a conditional and redundant homeostatic system involving all elements of peripheral, spinal, central and opiate modulation of pain transmission (WATKINS AND MAYER, 1982)." Several recent studies are contradictory to this opinion [58, 88, 132, 133, 134, 135, 160, 176, 196, 209, 237 and 238]. A more recent summary of the FDA is being expected with great interest.

PAIN AND WEATHER

HIPPOCRATE already placed a warning: "Do not use blood- letting, cauterisation or a knife" before a change in weather conditions. In certain regions (e.g. foehn, like in Innsbruck) surgeons even to-day follow this warning and do not perform certain operations before changing weather. Such changes also affect the sensation of pain. The author kept records of degree and duration of pain from 4.000 patients, as they described their complaints. It was noted that patients sensing changes in weather indicated a significantly higher degree and longer duration of their pain compared with those not having these sensations when weather is changing.

We tried to clarify these correlations by analysing retrospectively results of objective tests (such as blood chemistry, BSR) and comparing these with subjective data on a patient's sensing weather (or not). The most interesting, outstanding and unexpected results – the meaning of which we are unable to interpret as yet – were: Those patients sensing changes in weather conditions – especially on days of miserable feeling – show increased blood sedimentation rate, increased (mainly relatively, but sometimes also absolutely) number of lymphocytes, increased serum level of copper, decreased serum level of iron and in the pherogram there is marked increase in γ -fraction, mainly of α_2 - portion, while the albumin fraction is decreased. These changes sometimes are coupled with positive rheuma serology. All these findings may change day by day: on days the patients have most severe complaints, they are most marked while on days of well being results of objective tests may reach normal values. In looking for those weather factors possibly responsible for these changes we found increase in atmospheric pressure, of air temperature and of velocity of winds as well as decreased humidity. We started to call this combination a "weather mosaic" causing a corresponding "biologic mosaic", i.e. the changes in objective findings just mentioned.

We should like to state that there are objective findings showing a statistically significant and verified difference between patients sensitive to weather changes and those patients denying such sensitivity (we are unable to make any statements concerning normal persons). We therefore stipulate to ask all patients – as we do – their possible sensitivity to changing weather conditions. In case of a positive answer we ask that the form "F" (as shown in the appendix) be filled in during a whole month. This is meaningful only where a so-called "bio-weather" is calculated and published by meteorologists every day. If then for a certain period there is congruence of a patient's complaints (increasing) and the prevailing weather on exactly these days (same type) verification of this sensitivity to weather should be carried out by objective blood tests (see above): one should be carried out on a day of best and another on one of worst feeling of the patient. If however, there is no congruence of findings, other factors should be regarded.

The purpose of this objectivation of sensitivity to weather changes is two-fold: On the one hand, we treat all patients who are sensitive to weather changes much more intensively a priori than we treat those negating such sensitivity. On the other hand, in all instances where local meteorologic services are able to present a proper weather prognosis, we recognize a possibility to advise patients on what to do or what to avoid (especially evenings), if for the on-coming day the type of weather is being prognosticated which would bring on their usual reaction or complaints. In case of a correct prognosis it was possible in many more than 100 cases to at least markedly diminish [122] or totally prevent the complaints

sensing changes in
weather conditions

meteorologic
factors

OBJECTIVE CHANGES

OBJECTIVE CHANGES

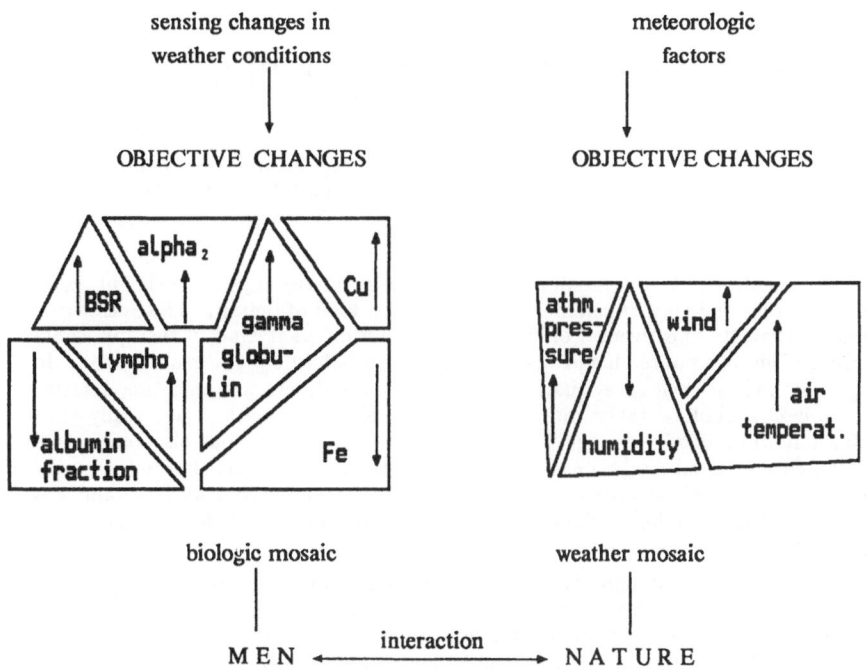

biologic mosaic

weather mosaic

M E N   ← interaction → N A T U R E

of the patients. Concludingly we advise physicians to pay more attention to the question of sensitivity to weather changes when dealing with the treatment of pain problems.

EVALUATION OF PAIN - PRACTICAL PROCEDURE

Often after a day, even after one hour you
laugh at a pain and do not feel the wound.
(Rückert; Weisheit des Brahmanen vol. 4, pg. 82)

Once a patient suffering from (chronic) pain has reported for treatment one should try to evaluate the degree and kind of pain before initiating any kind of therapy. This is essential to be able to evaluate effectiveness of therapy. Maybe a different kind of therapy would be required. This procedure should be adhered to also before electric pain control. In selecting a method for such an evaluation, the number of patients seen in a time interval (e.g. one week) should be taken into account. We selected a method not really time consuming but still objective enough to allow comparison with later controls: four factors were evaluated and numbers between 1 and 4 given to each of these. It originally was described as "pain profile scoring matrix" by PICAZA, SHEALY AND RAY [194] and was adapted to use at the "Ambulatorium Süd" under the name of "pain index". Explanation is given in table 3.

The practical procedure will follow this pattern: The referred patient presents his previous records. An exact case history is taken, generally and in respect to pain. For this purpose, specially prepared forms (see appendix) are used to save time and effort and to simplify documentation. An evaluation of the patient's pain is done as in table 3a. The cause of the pain should be determined: Whether the cause really is organic and whether it will be possible to interrupt pain conduction in the course of one or a few nerves. If the latter is the case (and this applies also to malignancies as causes) conduction in this pathway should be interrupted.

We shall now try to relate the experiences the author has collected on 34.000 patients in an effort to circumvent the impossibility of presenting all pain patients to pain clinics [ZIMMERMANN, 255]. It is hoped that a practicable alternative will be presented allowing readers to solve most problems by themselves after having some information on working principles of a pain clinic, basic information on pain projection as well as indications and contraindications of this specific kind of electro-therapy. These aspects shall be presented one after another before touching on special problems inherent in and peculiar for each nerve.

Factor	scaling of degree by numbers			
	1	2	3	4
Duration of pain in hours per day	0 - 6	6 - 12	12 -18	18 - 24
subjective degree of pain sensation	slight	moderate	severe	very servere unbearable
useful drugs	analgesics	sedatives	hypnotics	narcotics
psychic changes	mild	moderate	marked	severe (not cooperative)

Table 3a: Quantitative determination of "pain index" as sum of 4 single numbers. Possible values range from 4 - 16 [120].

36

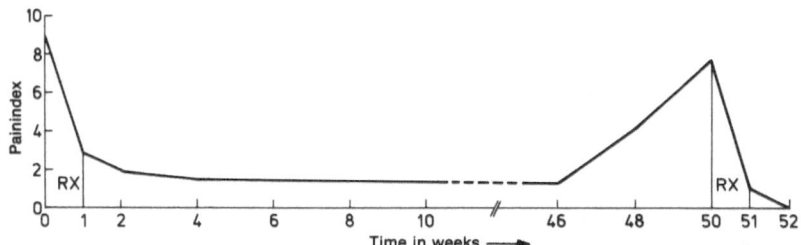

Fig. 11a: Longitudinal course of pain index in case of a recurrence of pain. From JENKNER [120].

How to approach this interruption – irrelevant from the cause of pain – may be seen in a schematic drawing (table 3b). Let it be pointed out, however, that the generalisation widely used and called "pain spiral" is an oversimplification which does not make any sense. The conditions prevailing in damage to nerves depend on the degree of damage (table 3c): slight damage will cause stimulation while severe damage will cause loss of function of a nerve (paralysis). For both instances there are three different possibilities depending on the type of nerve fiber affected by the damage: somato-sensory afferents, motor efferents or autonomic/sympathetic fibers. The various possibilities are to be seen from table 3c. It pays to remember that most nerves are mixed nerves and most commonly there will be a combination of the possibilities mentioned.

Influencing the function of a nerve may then be done most simply and without further damaging the nerve by the type of electrical impulses described under the section "optimal therapy".

Using the information provided by the scheme ("pain index") just mentioned makes it easy to evaluate the degree of pain in case of a recurrence. Such an event is presented graphically in fig. 11a. It is suggested to check the degree of pain once every month after the end of a therapeutic series. One should record the "pain index" one year after therapy to make a final evaluation, just as is shown in table 6 (pg. 54) Thus the practical advantage of an evaluation by pain index becomes evident.

ANAMNESIS

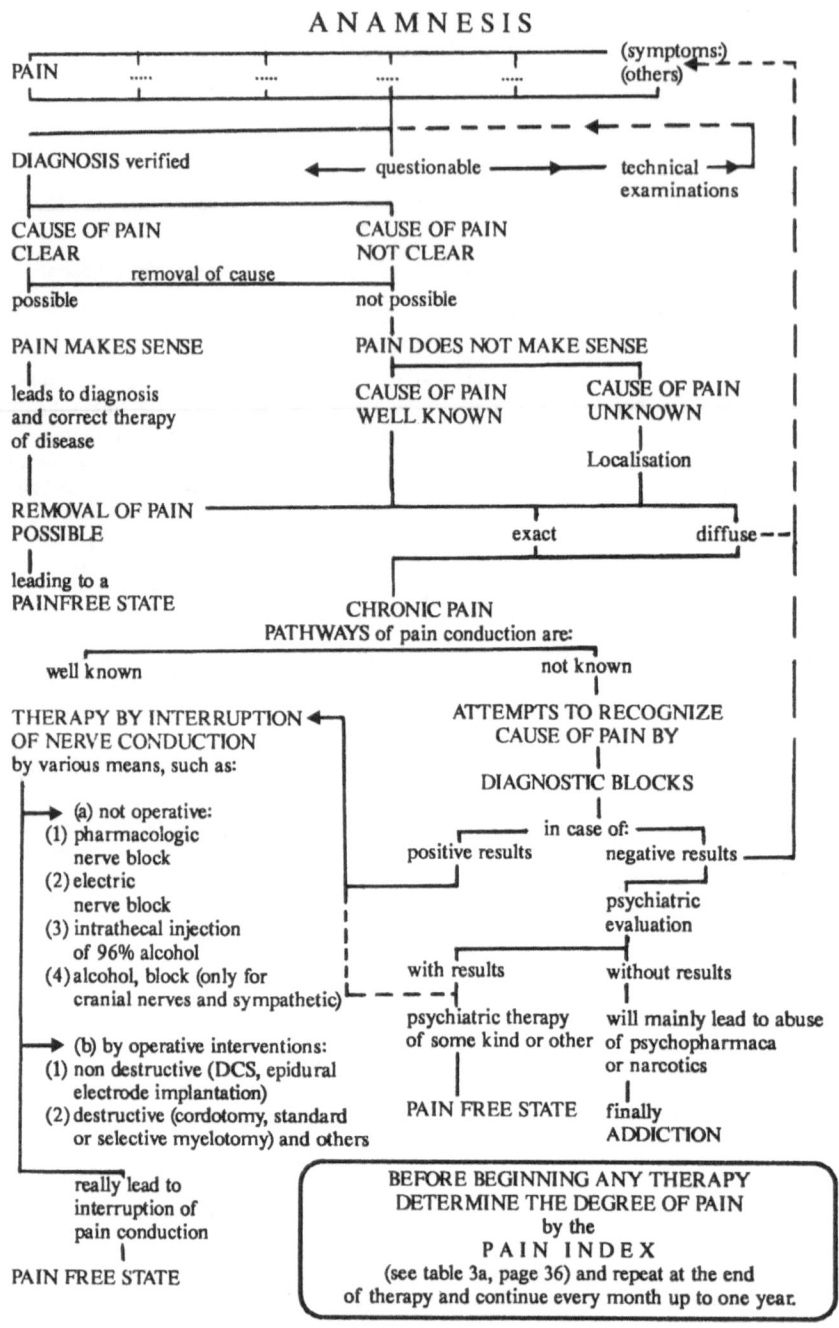

PAIN .... (symptoms:) (others)

DIAGNOSIS verified ← questionable → technical examinations

CAUSE OF PAIN CLEAR — removal of cause — CAUSE OF PAIN NOT CLEAR

possible   not possible

PAIN MAKES SENSE   PAIN DOES NOT MAKE SENSE

leads to diagnosis and correct therapy of disease

CAUSE OF PAIN WELL KNOWN   CAUSE OF PAIN UNKNOWN

Localisation

REMOVAL OF PAIN POSSIBLE

exact   diffuse

leading to a PAINFREE STATE

CHRONIC PAIN
PATHWAYS of pain conduction are:

well known   not known

THERAPY BY INTERRUPTION
OF NERVE CONDUCTION
by various means, such as:

ATTEMPTS TO RECOGNIZE
CAUSE OF PAIN BY

DIAGNOSTIC BLOCKS

in case of:

(a) not operative:
(1) pharmacologic
    nerve block
(2) electric
    nerve block
(3) intrathecal injection
    of 96% alcohol
(4) alcohol, block (only for
    cranial nerves and sympathetic)

positive results   negative results

psychiatric
evaluation

with results   without results

(b) by operative interventions:
(1) non destructive (DCS, epidural
    electrode implantation)
(2) destructive (cordotomy, standard
    or selective myelotomy) and others

psychiatric therapy
of some kind or other

will mainly lead to abuse
of psychopharmaca
or narcotics

PAIN FREE STATE

finally
ADDICTION

really lead to
interruption of
pain conduction

PAIN FREE STATE

BEFORE BEGINNING ANY THERAPY
DETERMINE THE DEGREE OF PAIN
by the
PAIN INDEX
(see table 3a, page 36) and repeat at the end
of therapy and continue every month up to one year.

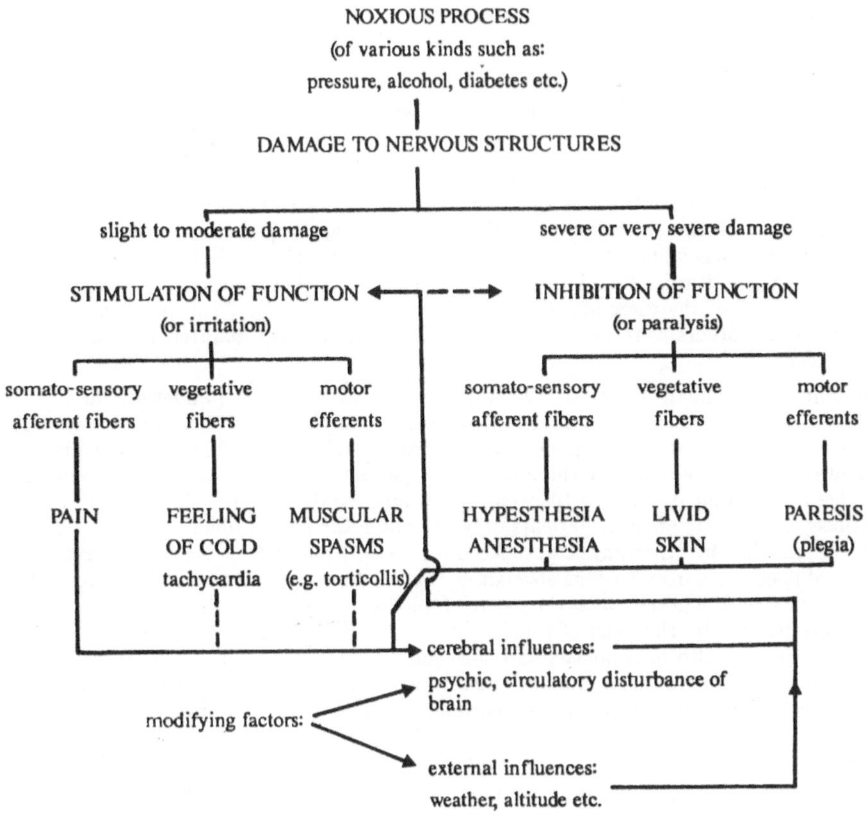

NOXIOUS PROCESS

(of various kinds such as:

pressure, alcohol, diabetes etc.)

DAMAGE TO NERVOUS STRUCTURES

slight to moderate damage   severe or very severe damage

STIMULATION OF FUNCTION ←--→ INHIBITION OF FUNCTION

(or irritation)   (or paralysis)

somato-sensory afferent fibers	vegetative fibers	motor efferents	somato-sensory afferent fibers	vegetative fibers	motor efferents
PAIN	FEELING OF COLD	MUSCULAR SPASMS	HYPESTHESIA ANESTHESIA	LIVID SKIN	PARESIS (plegia)
	tachycardia	(e.g. torticollis)			

cerebral influences:

psychic, circulatory disturbance of brain

modifying factors:

external influences:

weather, altitude etc.

Table 3c: Symptoms of slight or severe damage to nerves; differentiated for the three types of fibers and their interaction with the brain.

Table 3b: Sequence of approach to find adequate therapy for pain from the various symptoms of a disease.

PAIN CLINIC

Yes – Pain! Only you cause humans to become really human.
(Alphonse de Lamartine: Harmonies)

Ever since L. WHITE [249], a neurosurgeon at University of Washington, Seattle, Washington, USA grasped the impossibility in 1958 of diagnosing and treating all pain conditions by one medical speciality, "pain clinics" (in German "Schmerzambulanz") have been established. Soon after its foundation by L. WHITE the well known anesthesiologist J. J. BONICA took over the direction of the first pain clinic [20, 21]. After its plan of organisation, several others were founded and now there are several hundred in operation the world over.

The aim of such institutions is to have all various aspects of a patient's pain problem looked at by the respective specialists. Together, these physicians should arrive at a plan according to which particular pain could be treated effectively. The referring physician should be part of the team. Without a referral, patients could not just go to a pain clinic, because "their aunt or friend had been looked after so excellently" since a pain clinic is destined to take care of those patients only for whom the pain problem has not effectively been approached by other physicians. Accordingly, any pain clinic should have available a group of specialists from various specialties (at the Ambulatorium Süd there were 36) of whom one would, together with the referring physician, supervise the care of the patient. Seen internationally, the head of a pain clinic most often is either an anesthesiologist or a neurosurgeon. Also neurologists and specialists of physical medicine are directing such institutions sometimes. The director should have sufficient knowledge of neuroanatomy, neurophysiology, neuropathology and neuropharmacology. The development of pain clinics has shown a trend to disregard this interdisciplinary organisation. One should place more attention to the initial (i.e. really the best) plan of organisation of a pain clinic.

The following methods of treatment should be available at a pain clinic: Neural therapy, pharmacologic nerve blocks, manual therapy (or chiropractics), all kinds of physical therapy including massage, under water therapy, special exercises, ergotherapy, orthotic measures (such as improving prostheses etc.) and preparing patients for a job and giving advise for life conditions (e.g. for paraplegics), acupuncture, bio-feedback, psychological care or psychotherapy, as well as all types of TENS, especially the type proposed here using the type of impulse presented on page 21, on the application and results of which we shall have much information for the reader. Furthermore, all neurosurgical interventions for pain relief from epidural implants to stereotactic procedures should be possible. A properly equipped and staffed pain clinic should be able to provide computer documentation, statis-

---

Fig.12: Brief survey on the number of patients seen between 1976 and 1989 at the pain clinic of the "Ambulatorium Süd" of the Vienna Sick Fund (A-1100 Vienna, Wienerbergstrasse 13) and treated: number of new patients (upper diagram, lower columns). The division of lower diagram indicates change of treatment modalities over the years as observed; only most frequently used items are depicted, rare kinds of therapy such as e.g. neurosurgical interventions or bio-feed-back are not shown.

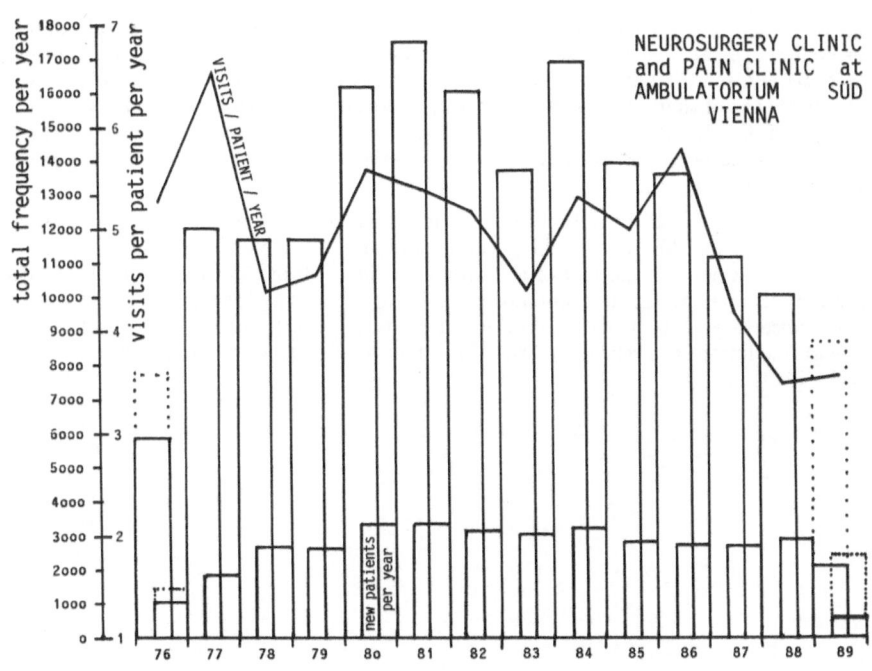

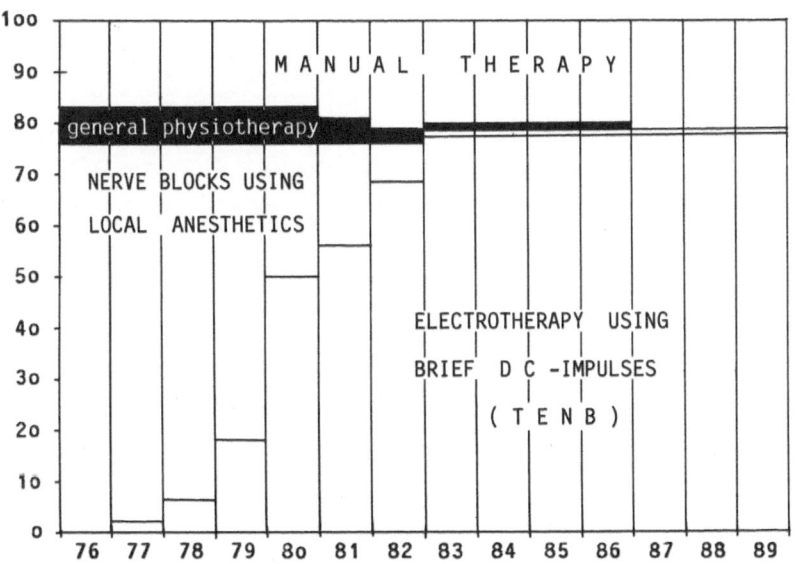

%-fraction of various forms of therapy for treating pain

tical evaluation of all pertinent data to carry out research projects and provide monthly pain conferences as well as other means of continuing medical education for interested physicians and related personnel. The number of patients handled over the years and the changing means of therapy occurring at "Ambulatorium Süd" pain clinic, Vienna becomes evident from fig. 12. Rare forms of therapy have not been included, such as neurosurgical operations or bio-feedback; the number of new patients and visits per patient per year are however included.

At the pain clinic which the author headed, one tried to adhere to the principles presented and arrive at a feasible plan, solving the average patient's pain problem and providing a plan for treatment at the end of a morning session. More time was necessary if time-consuming studies or timed examinations such as Tc-scans, MRI or laboratory studies became necessary as part of the work-up. How to approach difficult pain problems and arrive at a solution in a relatively brief time shall be discussed in the next chapter.

PAIN PROJECTION, SEGMENTAL ACCESS
AND SCLEROTOMES

In any kind of irritation of a somatic afferent nerve somewhere along its course there is a projection of all sensations (including pain) to that part of the body where the irritated nerve is coming from. A projection of the subjective pain sensation derived from an organ is very specific for each organ as are the so-called "Head's zones". Knowledge of these conditions is very helpful in recognizing possible causes of a pain, whose origin need not always be situated in the periphery of the nerve. A very suitable object for demonstration of these circumstances is phantom pain. Here, irritation of a nerve (or some specific pathway) will cause a pain projection to parts of the body (e.g. lower leg) which are not existing any more. This example may also be used to state that pain will only be sensed if the relation of afferent somatic and sympathetic influx to the brain is disturbed. After restoring the normal relation, the patient will be free of pain.

Which segmental fibers or nerves (somatic and sympathetic) carry impulses from or to specific organs may be seen in table 4 (changed after RAUBER - KOPSCH, from JENKNER [118]. This table is supplemented by data on the segmental access to certain muscles and reflexes (table 5; extended from JENKNER [118]). There are several models describing the segmental division of the skin. We advise using the one arrived at and used by the department of neurosurgery of the University of Zürich (fig. 13; with kind permission of this department).

This tabulary presentation however should not tempt to treat any pain only by blocking nerves and forgetting the underlying disease, for which this pain may be characteristic. These tables are not to be regarded as complete information. Rather they should be regarded as an adjunct once one has decided to treat a particular pain via nerve blocks. The same applies to the four following figures supplying information on segmental (or peripheral) supply of bones, ligaments and the periosteum with muscle attachments (fig. 14 - 17).

Since many physicians sometimes are reluctant to use pharmacologic nerve blocks by local anesthetics, fearing side effects (which however may be circumvented or treated easiliy; JENKNER [118]), the previously described and proven effect of electric impulses of a special kind, resulting in true nerve blocks, is obtaining great significance.

It should be recalled that pain is conducted not only by somatic nerves. Conduction also takes place via true sympathetic nerves and fibers traversing sympathetic ganglia without being part of a synapse there. This is the reason these fibers may be blocked simultaneously with the ganglion, even though they are not sympathetic fibers. (GUILLAUME et al. [86]). It needs to be emphasized that the assignment of medullary segments to organs unfortunately is not uniform; from author to author there are differences. We therefore have made a selection for the presentation in the table. Using the average or central segment given one certainly will be acting correctly. In the use of electric impulses for blocking nerves, there are inexactitudes due to field distribution variations. These may be an advantage: If somatic and sympathetic nerves both need to be blocked alternatively, which is indicated at times (e.g. in cancer or post zoster pain) they are blocked simultaneously by this variation. Another example is the interruption of lumbar sympathetic fibers where aberrant fibers, too small to be seen by a surgeon, may even disrupt efficacy of an opera-

Table 4:

Survey of Pain Projection

Organ	Pain projection	Pain conduction via segmental nerves	Origin of preganglionic sympathetic fibers
Head and neck			via stellate ganglion
Meninges	Scalp	Nucl. sens. N.V.* IX, X & XII	Th 1 - Th 2 (3)
Eye	Eye socket forehead	Nucl. sens. N.V. (first branch)	Th 1 - Th 3 (4)
Tear glands	Eye socket	Nucl. tract. solit. N. VII & IX	Th 1, Th 2
Parotid gland	Parotid region	Nucl. tract. solit. N. V via N. VII + IX	Th 1, Th 2
Salivatory glands	Submandibular region	Nucl. tract. solit. N. ling. via N. VII and geniculate ganglion	Th 1, Th 2
Thyroid gland	Ventral part of neck	C 2 - C 4 Th 1, Th 2	Th 1, Th 2
Larynx	Throat and ventral part of neck	N. laryng, sup. ganglion	Th 2 - Th 7
Thorax			
Trachea, bronchi	About the sternum	Th 2 - Th 7	Th 2 - Th 7
Lung parenchyma	Insensitive for pain	Insensitive for pain	Th 2 -Th 7
Parietal pleura in area of:			
Shoulder	Shoulder	C 3 - C 5	
Supraclavicular	Supraclavicular (brachial plexus)	C 8 - Th 1	
Intercostal	Intercostal nerves	Th 1 to Th 12	
Heart	Precordium and left (right) arm	Th 1 - Th 4 (5)	Th 1 - Th 4 (5)
Thoracic aorta	Upper half of thorax and neck	Th 1 - Th 5 (6)	Th 1 - Th 5
Abdominal aorta	Lower half of thorax and abdomen	Th 6 - Th 12	Th 6 - L 2
Esophagus Upper half	Mid-sternum	Th 5 - Th 8	Th 2 - Th 5
Lower half			Th 5 - Th 8

* Roman numerals refer to cranial nerves.

Table 4 (cont.):

Survey of Pain Projection

Organ	Pain projection	Pain conduction via segmental nerves	Origin of preganglionic sympathetic fibers
Abdomen			
Stomach	Epigastric and interscapular region	Th(6)7, Th8(9)	Th(5)6 - Th 10 (11)
Liver and gall bladder	Right hypochon-drium	Th(5)6 - Th 8(9) and phrenic nerve	Th 6 - Th 11 right
Pancreas	Epigastrium, lower part of sternum, midline on back in area of 10th and 11th rib	Th(5)6 - Th 10(11) and vagus nerve celiac ganglion	Th 5 - Th 11 left
Spleen	Left hypogastric area	Th 6 - th 8	Th 6 - Th 8
Small intestine: Duodenum	Epigastrum and	Th(5)6 - Th 7(8*)	Th 6 - Th 11
Jejunum and Ileum	Umbilical area	Th 9 - Th 11	
Large intestine: Cecum and ascending colon	Suprapubic area	Th 9 - Th 11	Th 8 - Th 11 right
Appendix	Right lower quadrant	Th 10 - Th 11 (-L 1)	Th 8 - Th 11 right
Descending colon and sigmoid	Deep pelvic area and anus	L 1 and L 2 S 2 - S 4	Th 11 - Th 12 L 1 - L 4 left
Adrenal gland	None	None	Th 6 - L 2 u**
Kidney	Hip and groin	Th 10 - L 2	Th 10 - L 1(2) u
Ureter	From back to groin	Th 11 - L 2	Th 11 - L 1(2) u
Urinary bladder: Fundus	Suprapubic area	Th 11 - L 1	L 1 - L 2 (hypogastric nerve)
Neck	Perineal and anal region	S 2 - S 4	
Testicles	Testes	Th 10	Th 10 - L 1
Prostate gland	Perineal area and lower back	Th 10, Th 11 S 2 - S 4	Th 10 - L 1
Ovaries and tubes	Both lower quadrants	Th 10	Th 6 - L 2
Uterus	Perineum, lower pelvic area	Th 10 - L 1, S 2 - S 4	Th 6 - L 2
Female external genitalia	Perineum and local	S 2 - S 4	L 1 - L 2
Extremities			
Blood vessels, sweat glands, hair follicles etc. of: Upper extremity	Local on skin	C 5 - Th 1	Th 2 - Th 8(9)
Trunk	Local on skin	Th 1 - Th 12	Th 1 - Th 12
Lower extremity	Local on skin	L 4 - S 3	Th 10 - L 3

* Oral half right, aboral half left and vagus nerve,
** unilateral

tion, but will be interrupted by the lack of exactitude of field distribution in the depth of the sympathetic chain. The final result may even be better than after surgery.

The figs 14 -17, supplementing the information from the tables, are presented with kind permission of INMAN and SAUNDERS [107]. Their great value remains in the fact that with one glance and a reliable description of the pain by a patient one is able to differentiate immediately between segmental and peripheral nervous causes of a pain distribution. This is of utmost importance in shoulder pain, clavicular pain and pain in cancer patients where primary or secondary lesions are penetrating bones or in joint pain.

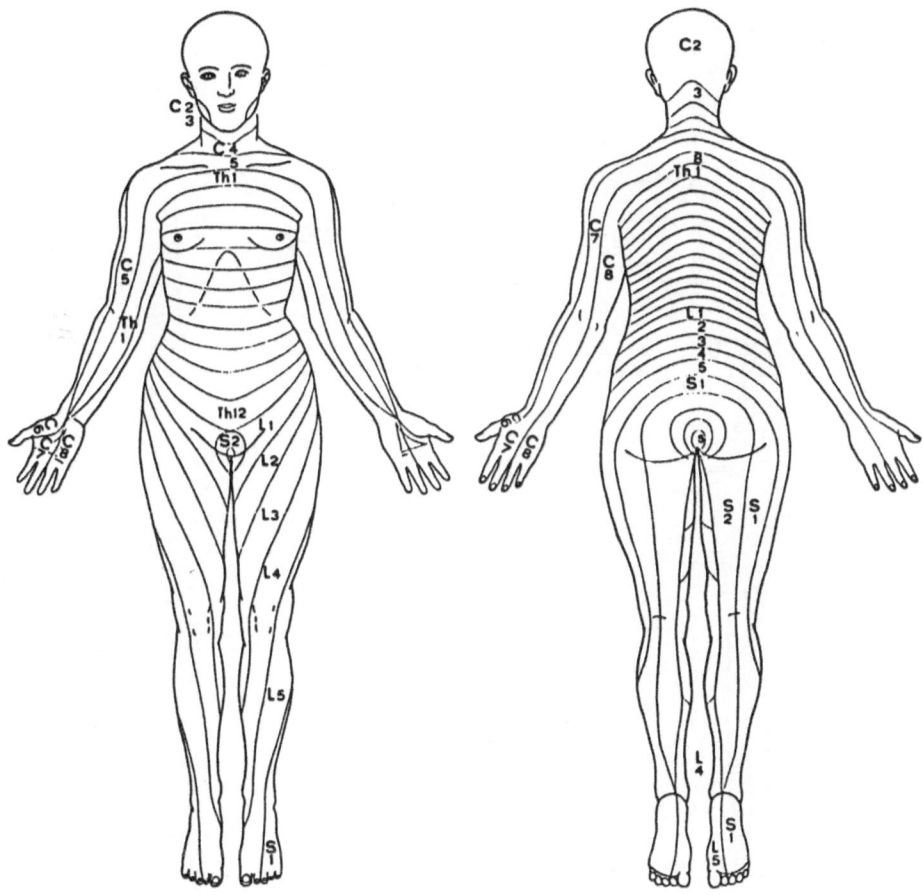

Fig. 13:   Segmental sensory scheme arrived at and used at "Department of Neurosurgery, University of Zurich". With kind permission of this department and the owner of the copyright, Msg. J.R. Geigy S.A., Basle, Switzerland. Easy to remember are the following levels: C - 2 bordering first branch of trigeminal = hair border on forehead; C - 3 towards 3rd branch of trigeminal = Ear lobe to lower jaw / neck; C - 4 towards Th - 1 = clavicle; C - 6 = thumb; C - 8 = little finger; Th - 4 = nipples; Th - 10 = navel; Th - 12 to lumbar segments = inguinal ligament / fold; L - 5 = great toe; S - 1 = small toe (dorsal, bilateral)

47

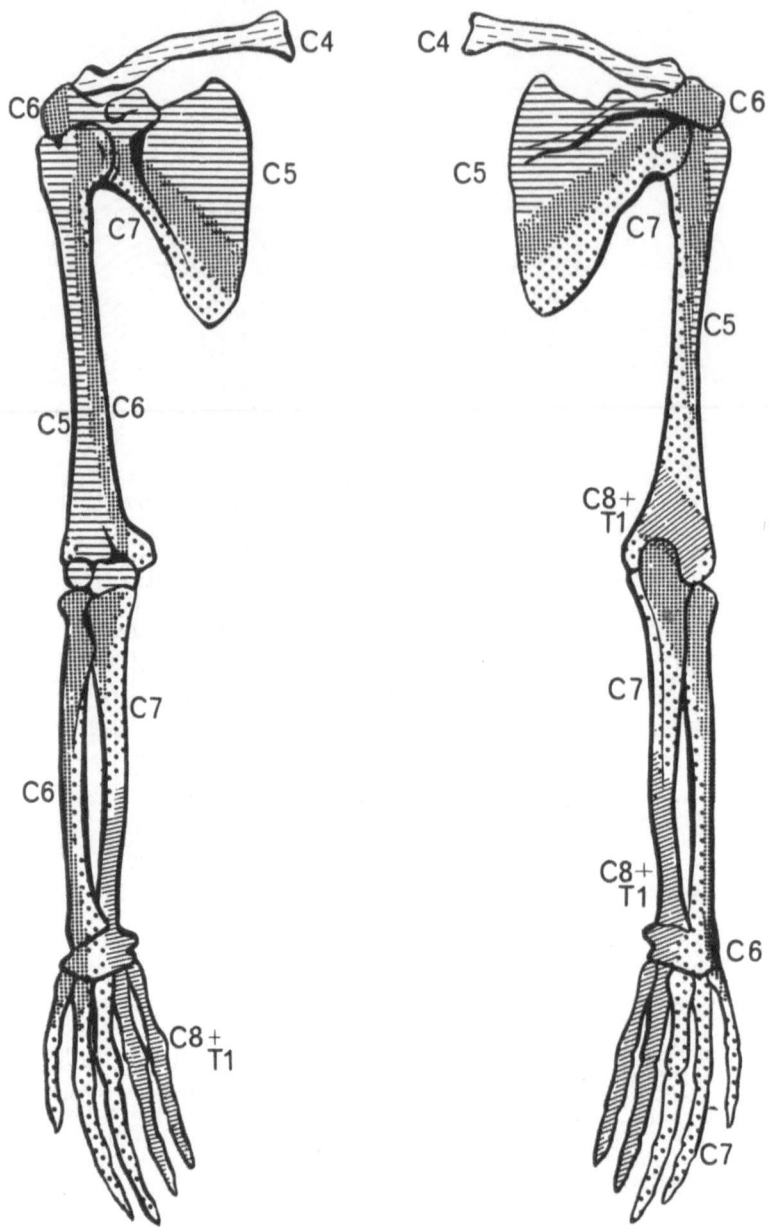

Fig. 14: Sensory scheme of bones of arm for medullary segments.

48

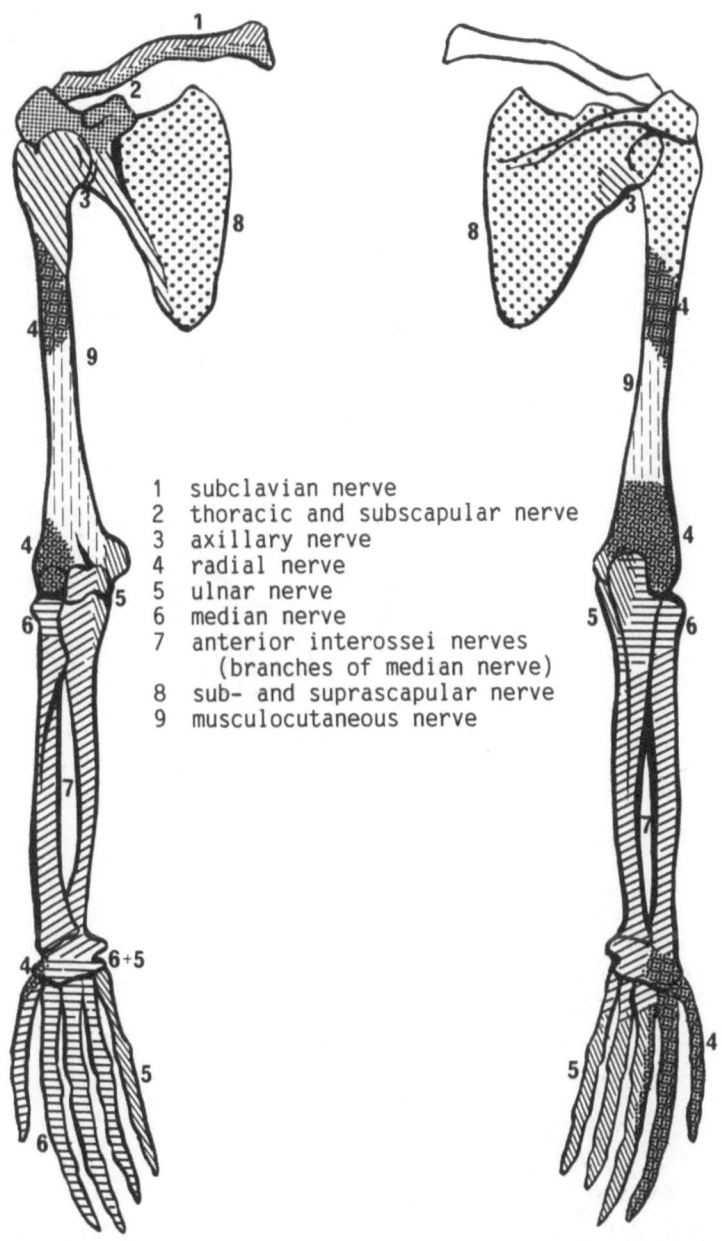

1 subclavian nerve
2 thoracic and subscapular nerve
3 axillary nerve
4 radial nerve
5 ulnar nerve
6 median nerve
7 anterior interossei nerves
   (branches of median nerve)
8 sub- and suprascapular nerve
9 musculocutaneous nerve

Fig. 15:   Sensory scheme of bones of arm for peripheral nerves.

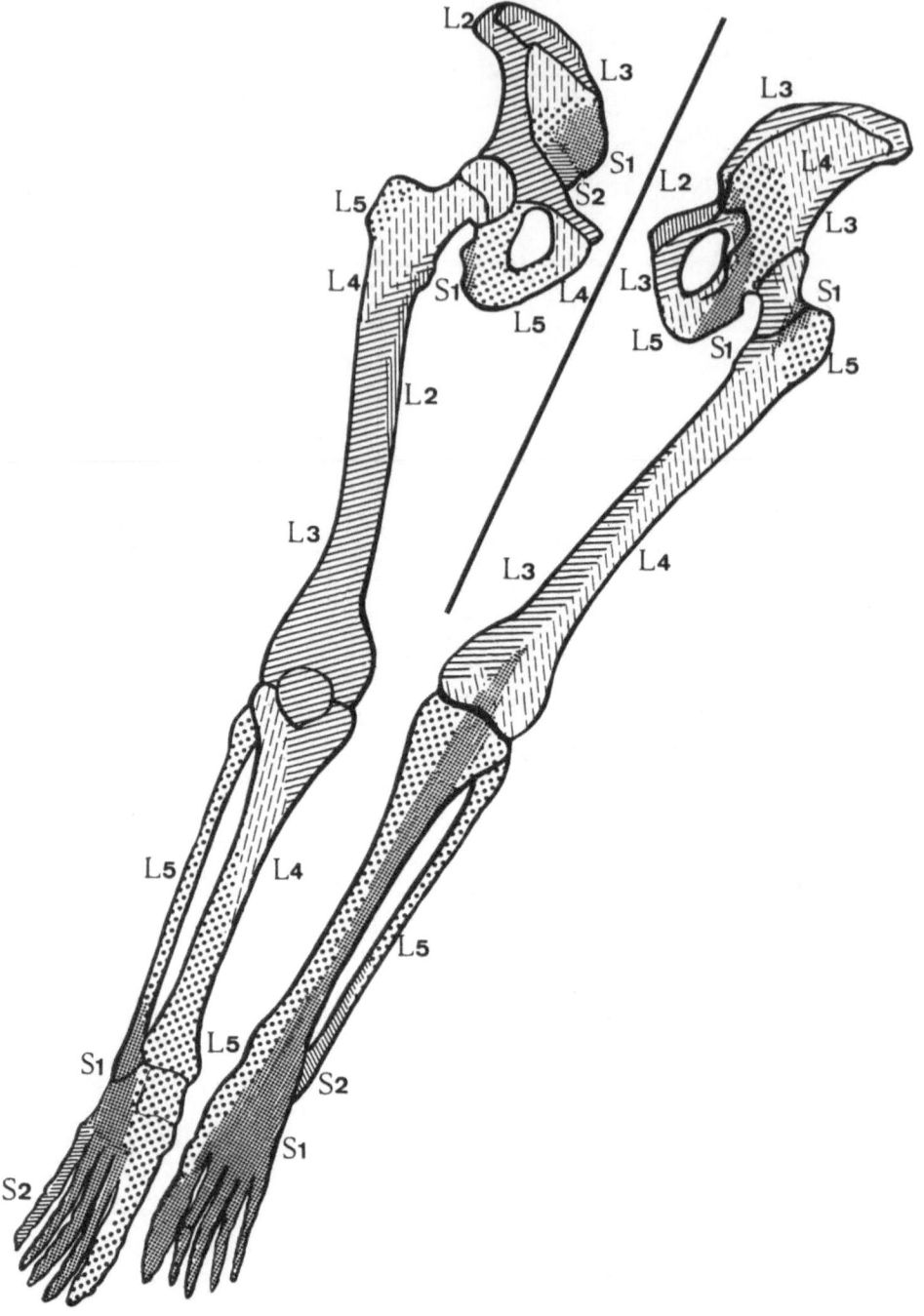

Fig. 16:   Sensory scheme of bones of legs for medullary segments.

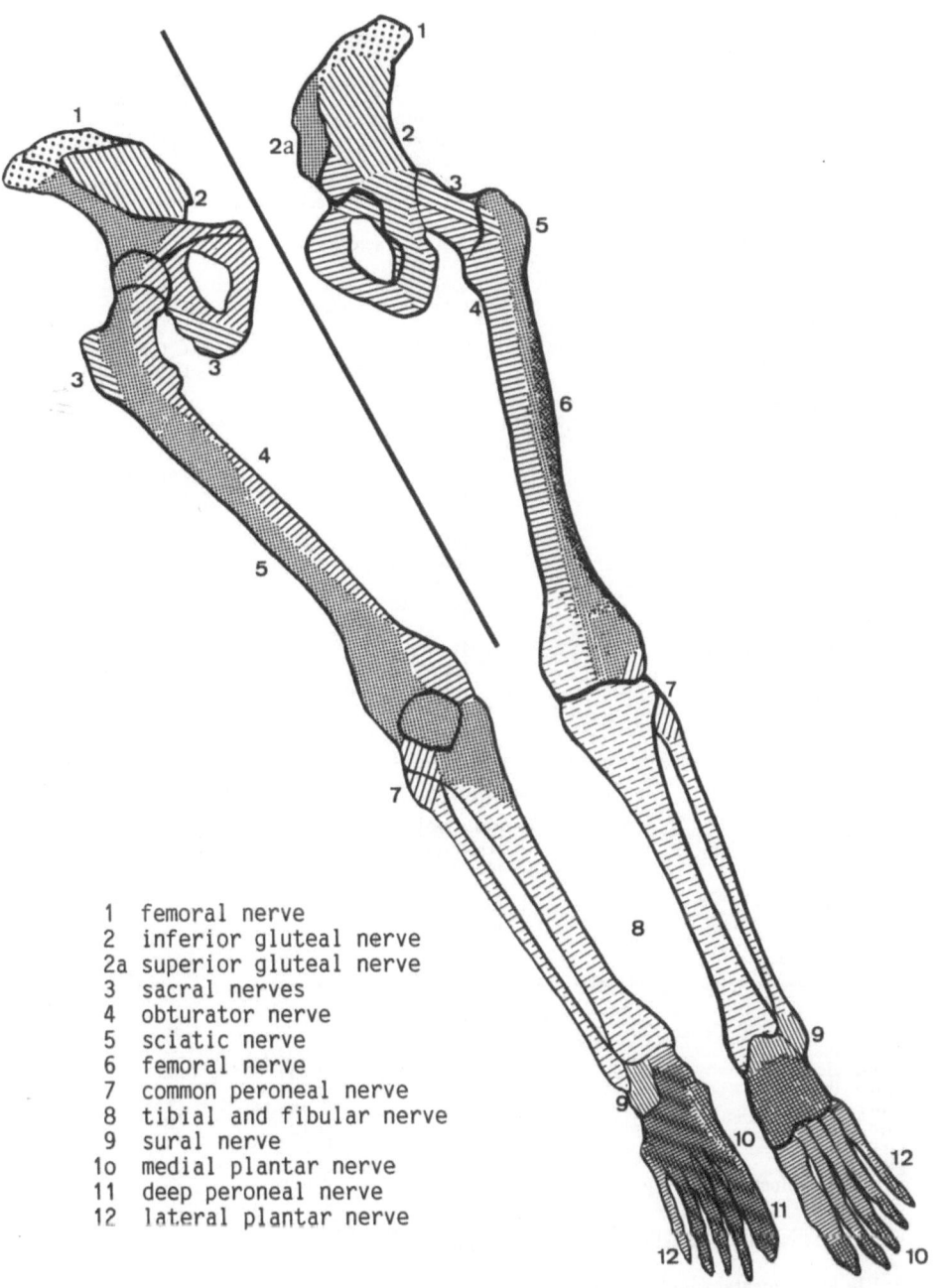

1   femoral nerve
2   inferior gluteal nerve
2a superior gluteal nerve
3   sacral nerves
4   obturator nerve
5   sciatic nerve
6   femoral nerve
7   common peroneal nerve
8   tibial and fibular nerve
9   sural nerve
1o medial plantar nerve
11 deep peroneal nerve
12 lateral plantar nerve

Fig. 17:   Sensory scheme of bones of legs for peripheral nerves.

Segmental supply of Muscles

muscle		nerve	segment
masseter, temporal		trigeminal	medulla
soft palate, pharynx		glossophar., vagus	medulla
tongue		hypoglossal	medulla
deep neck muscles		cervical	C 1 to 4
sternocleidomastoid		accessory and cervical	medulla and C 1-3
trapezoid		accessory and cervical	medulla and C 2-4
diaphragm		phrenic	C 3-5
major pectoral		ant. thoracic	C 5-8, D 1
deltoid	**C-5**	axillary	C 5-8
anterior serrated		long thoracic	C 5-7
biceps	**C-6**	musculocutaneous	C 5+6
brachioradial		radial	C 5+6
triceps	**C-7**	radial	C 6-8
extensors (wrist, fingers)		radial/ulnar	C 5-8
flexors of fingers		median/ulnar	C 7-D 1
muscles of thenar	**C-7**		
muscles of hypothenar	**C-8**		
intercostal		intercostal	D 1-12
abdominal wall muscles		intercostal	D 6-12
muscles of back		spinal	D 1-S 5
iliopsoas		femoral	L 1-4
max. gluteal		inferior gluteal	L 5-S 2
medial gluteus		inferior gluteal	L 4-S 1
quadriceps of femur	**L-3**	femoral	L 2-4
adductor	**L-3**	obturator	L 2-4
biceps, semimembran. and semitendinosus		sciatic	L 5-S 2
soleus, gastrocnemius		tibial	L 5-S 2(3)
peroneal group	**L-4**	common peroneal	L 4 - S 1(2)
triceps surae	**S-1**		
perineal and sphincters		pudendal	S 3-5

Segmental supply of Reflexes

Tendon reflexes

lower jaw jerk	trigeminal nerve	medulla
BTR	musculotaneous	C 5+6
TTR	radial nerve	C 6+7
knee jerk	femoral nerve	L 2-4
ankle jerk	tibial nerve	S 1-2
radio-periosteal		C 8-D 1

Superficial reflexes

cilio-spinal	C 8-D 1
epigastric	D 7+8
upper abdominal	D 8+9
lower abdominal	D 10+11
suprapubic	D 12
cremasteric	L 1+2
gluteal	L 4+5
plantar (Babinski)	S 1+2
anal reflex	S 5

(C = cervical; D = dorsal or thoracic; L = lumbar; S = sacral)

Table 5: Segmental supply of muscles (via their nerves) and of tendon and superficial reflexes. **Bold** face types indicate segments for which the respective muscle is representative. Greatly extended after JENKNER [118].

INDICATIONS

Where it hurts, there you touch.
(German saying)

If the directions given so far have allowed the reader to establish the cause of some segmental pain, the somatic or sympathetic nerve should be known. It then becomes necessary to place the anode over this nerve to treat the pain. In conditions where such segmental allocation is impossible (e.g. joint pain), it is advisable to select an impulse with a higher DC-component whenever the apparatus provides such modalities of current. If this is not possible, the duration of a single pulse should be doubled. Keep the size of electrodes as different as for treating nerves and use the small electrode as the anode over the most painful area (e.g. in gonarthrosis, figs. 93 and 94). The duration of the therapeutic session should be 20 minutes. This subtle difference in type of current should be noted also when regarding table 6. Higher DC-component is equal to the FM- selection on some apparatus (e.g. Electroblock, Relaxette; see table 2).

The only pain in a joint not to be treated in this manner is coxarthrotic pain, if the hip joint had not been operated previously. On how to treat this pain, turn to the section on obturator nerve. The various types of pain which were found to respond to this type of therapy at the pain clinic of the "Ambulatorium Süd", how many patients were seen and with what results, may be seen in table 6.

	DIAGNOSES	Case Numbers	procent of group	% figures of all patients	success RATE
A.	NEURALGIAS total	9420	100	52,7	75,2
	Cervical Neuralgias	6461	69,9	36,8	74
	post-herpetic neuralgias	619	6,7	3,5	70
	trigeminal neuralgia	603	6,5	3,4	84
	low back pain (lumbalgia)	343	3,7	2,0	68
	intercostal neuralgia	297	3,2	1,7	76
	phantom pain	151	1,6	0,9	69
	sciatica	116	1,3		64
	sacral neuralgia	64	0,7		55
	ulnar neuralgia	26			84
	neuralgias, other	740	8,0		72
B.	SYMPATHETIC DYSTROPHIES, CIRCULATORY DISTURBANCES; total	1801	100	10,3	71,5
	circulatory disturbances, brain	728	40,4	4,2	73
	- of upper extremity	392	21,8	2,2	71
	- of lower extremity	321	17,8	1,8	69
	hyperhidrosis	178	9,9	1,0	62
	SUDECK's atrophy	64	3,6		74
	lymphedema of arm, post ablatio	35	1,9		97
	pancreatitis, subacute & chronic	64	3,6		77
	Menière's syndrome and disease	11			84
	precordial pain	8			82
C.	ARTICULAR PAIN, total	4436	100	25,3	76,7
	sacro-iliac pain	2210	49,8	12,6	84
	periarthropathy, humero-scapular	847	19,1	4,8	72
	epicondylitis, radial and ulnar	718	16,2	4,1	68
	coxarthrosis (via obturator nerve!)	182	4,1	1,0	66
	gonarthrosis	123	2,8		64
	peritendinoses	126	2,8		66
	omarthrosis	15			68
	talo-calcaneal pain	11			66
	arthroses, other	204	4,6	1,2	67
D.	BENIGN DISEASES, Various, total	651	100	3,7	
	vertebral fractures, non recent	515	79,1	2,9	84
	various neurological disorders	68	10,4		88
	undiagnosed complaints, periarticular	33	5,1		71
	spasm of ureter, acute pain	26	4,0		78
E.	BENIGN TUMORS, total	18			50-93
F.	MALIGNOMAS, total	1189	100	6,8	60
	primary lesions of: bladder	26	2,2		45
	stomach	20	1,7		97
	testes	13	1,1		23
	-, other	31	2,6		68
	secondary lesions of -mammae	418	35,2	2,4	72
	-bronchi	259	21,8	1,5	58
	-recti, -sigmoid	176	14,8	1,0	49
	-prostate	42	3,5		64
	-uterus	18	1,5		60
	-epithelial tumors	26	2,2		66
	unclassified primary tumors	87	7,3		68
	other occult primary neoplasms	73	6,1		49-67
	pain reduction in first 2000 patients	2000			65
	pain reduction in last 1537 patients	1537			92
	TOTAL NUMBER OF PATIENTS	17537			

Table 6: Listing of diagnoses (in groups) with case numbers observed at the pain clinic Ambulatorium Süd; giving relative frequencies of diagnoses and the rate of pain reduction noted one year after therapy.

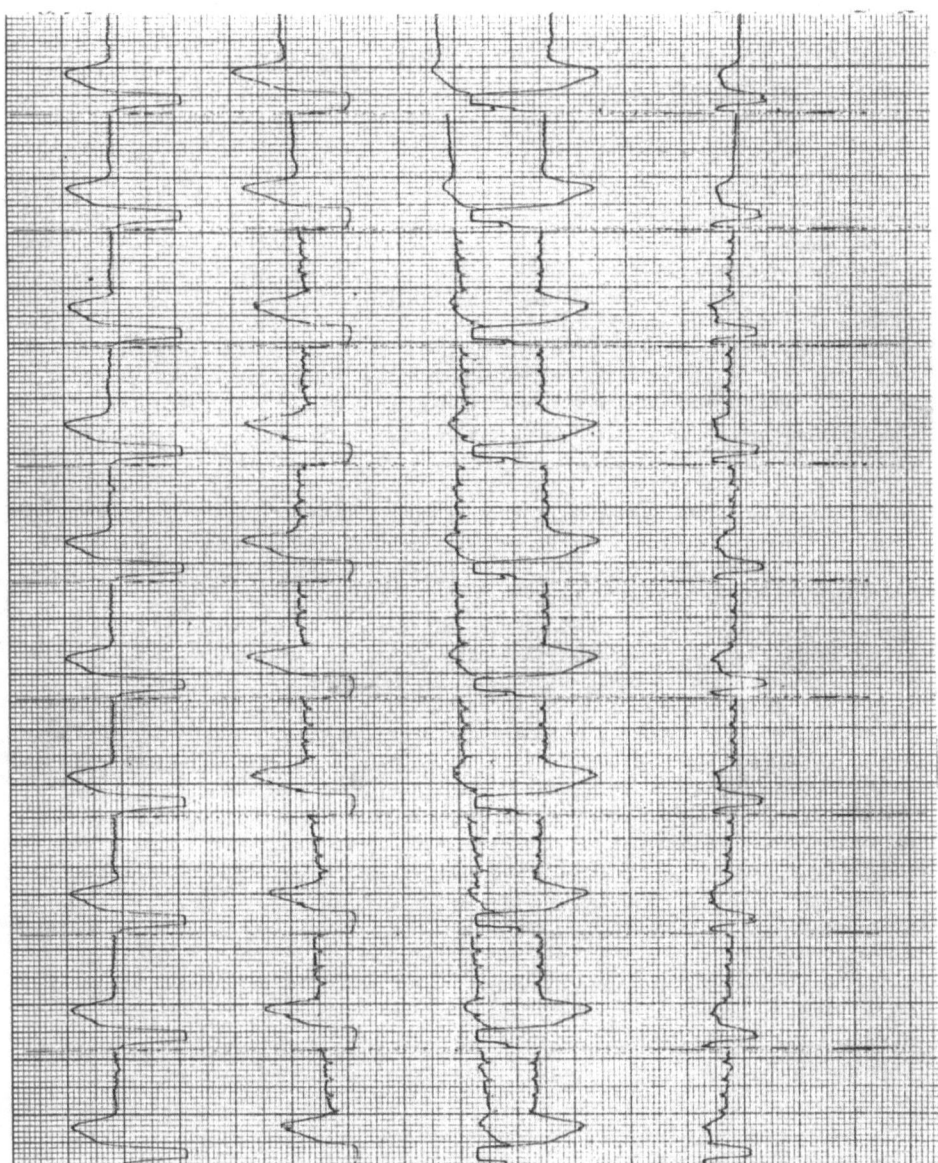

Fig. 18: Part of a continuous registration of electrocardiogram in a patient using a demand pace maker (implant) who requires electric pain control. ECG- monitoring is a conditio-sine-qua-non and therapy may only be carried out as long as pace maker signs of ECG are present (such as is the case in this recording). From JENKNER [118].

CONTRAINDICATION, COMPLICATIONS
LACK OF EFFICACY

Patients with an implanted demand-type pacemaker must not be subjected to this type of therapy since the short impulses will be regarded as a heart action by the pacemaker and it will quit functioning. Only if electrode positions are far away from the site of implant may an attempt be made to use electric nerve blocking as explained here; however, continuous monitoring of the electrocardiogram by a registered nurse/trained health care provider is required. The attendant should turn off the stimulation unit if at least two pacemaker complexes fail to show on the ECG. If the action of the pacemaker is continuing, as e.g. in the patient of fig. 18 (part of a continous recording), therapy may continue under the conditions of monitoring and supervision as indicated.

The only complications observed in our large experience were those which are well known also from pharmacologic blocking; structures very close to the nerve being treated are likewise influenced by the electric field. This is the case e.g. when blocking the stellate ganglion: a temporary hoarse voice may ensue as a consequence of blocking the recurrent nerve; or an increase in heart frequency may be seen as a consequence of vagal block when treating the superior cervical ganglion electrically. Such influences may not be avoided. They require informing the patient.

Metallic implantates or foreign bodies do not constitute a contraindication for this special form of electric impulses (page 21). The very short duration of a single impulse together with a frequency of 35 cps makes the pause between two consecutive pulses very long (more than 100 times the duration of a single impulse). This suffices to reverse all possible changes of tissue (such as polarization; which of course is occuring) to normal; each consecutive impulse therefore meets with normal conditions of tissue. Because of the small amount of energy involved no increase in temperature of such implants is possible (studied by animal experimentation and using more than 10 times the current possibly applied in man therapeutically).

Using various electrode pastes as contact media between electrodes and skin may at times cause allergic reactions. Tape sometimes also has such effects. While using karaya as contact medium and stick-on substance over a 5-year period, we never have observed allergic phenomena. Should patients mention that they do not tolerate electric current, no effort to persuade them to electrotherapy need be made. These patients are neurotics and any type of electric therapy would only bring increased difficulties and more pain.

The first check up is done after the fifth therapeutic session of 20 minutes. If efficacy be insufficient, the considerations should be directed to the following possibilities:

(1) Was the cause of the pain correctly recognized? (2) or was a wrong medullary segment or the incorrect nerve selected as site of therapy? (3) or is the patient taking drugs of the group of dopamine antagonists? If so, one should determine which is more essential, drug therapy or pain reduction. If the latter is regarded as first priority, the drug (e.g. metoclopramid) has to be discontinued. Lastly (4) if none of these conditions prevail and no reason for the lacking effect be found, one ought to try medication of a dopamine agonist such as a small dose of madopar or sinemet (25 or 50 mg) one hour before starting electrotherapy. After 5 such combination treatments, there should be a new evaluation of the effect. One may be able to sufficiently reduce pain in about 50% of the previous failures.

Should this not be the case, more difficult and selective measures have to be undertaken such as direct stimulation of the relevant nerve or root using an electrically insulated needle (exept for the tip) which greatly increases field density about the nerve. If such an epidural root stimulation has the desired effect, an epidural electrode may be implanted under local anesthesia by a neurosurgeon. If a proper selection of a surgical intervention is made and executed conscientiously, these measures usually are rather effective. There are several operations which are designated as functional neurosurgical procedures, but space forbids delving further into these procedures. One should know, however, that they exist.

Use of needle electrodes as anodes:

For those readers wanting to use needles – and feeling able and being able to insert these properly – to the close vincinity of those nerves that should be subjeted to a very dense electric field for better effects, a few considerations are in order. One must face the possibility to damage nerves if characteristics of the current (or impulses) applied are too high. The important data here are: exposed tip area, current density and charge density at a peak current/pulse of 8 mA (upper limit the author advises).

Furthermore, any metallic anode will release metal ions into the electrolyte solution (i.e. body fluids). Hence, metallic ions will definitely be left in the body after an electric treatment. The amount of these is being determined by the following equation:

$$\frac{\text{Oxidation rate of anode metal}}{\text{(in gram per coulomb)}} = \frac{\text{(molecular weight in grams)}}{\text{(Avogadro's number)(valence)(charge of electron)}}$$

a resulting figure which then has to be multiplied by the total number of coulombs from (in our case) monophasic square waves:

$$\text{(peak current)(pulse width)(frequency)(treatment time)}$$
$$= 8 \times 0{,}1 \ . \ 30 \ . \ 10 = 21\ 000 \ \text{microcoulomb}$$

For a stainless steel needle, this would amount to 4,05 microgram iron / treatment assuming stainless steel to contain roughly 70 % iron. The exact composition of AISI (SAE 30302) stainless steel is given as 69 % iron, 18 % chromium. 9 % nickel, 2 % manganese, 1 % silicon, 0,15 % carbon, 0,045 % phosphorus and 0.03 % sulfur. Proportional amounts of the metals in this list would make up the 4,05 micrograms above. Average urinary iron excretion equals 100 (40 – 150) μgrams per day. The question of gases being liberated at the site of the anode (oxygene, hydrogene and others) may be disregarded since the time beetween two impulses (more than 28 milliseconds or more than the 280 fold time of a pulse width) suffices for the gases to dissolve and be carried away with circulation. The figures of exposed tip area etc. are given in the adjoining table for two of the most commonly used needles.

needle designation	type	gauge	outer diameter	tip area	current density at 8 mA
TOP disposable needle surgical pack (see pg. 26)	hollow	23	0,023" 0,58 mm	0,64 mm^2	12,5 mA / mm^2
STIMEX No. 15150 Becton Dickinson	hollow	22	0,025" 0,62 mm^2	1,28 mm^2	6,25 mA / mm^2

energy /mm²	voltage at 8 mA	resistance at 8 mA in saline physiol.	bubbles at anode	needle
2,5 uC	4 V	500 Ohms	yes	TOP
1,25 uC	3,4 V	425 Ohms	no	STIMEX

To insure that no more than 8 mA are flowing instruments should be used which are equipped with a meter to show actual maximum current of a single pulse and these instruments must be equipped with a current limiting device. No such stimulators are currently available commercially: users must have their stimulators equipped with such meters or must use a stimulator-tester (also not available commercially) allowing to use all instruments with proper care. The possible damage to nerves rests with the users and not with the manufacturers of stimulators. This remark is necessary because of possible liability claims.

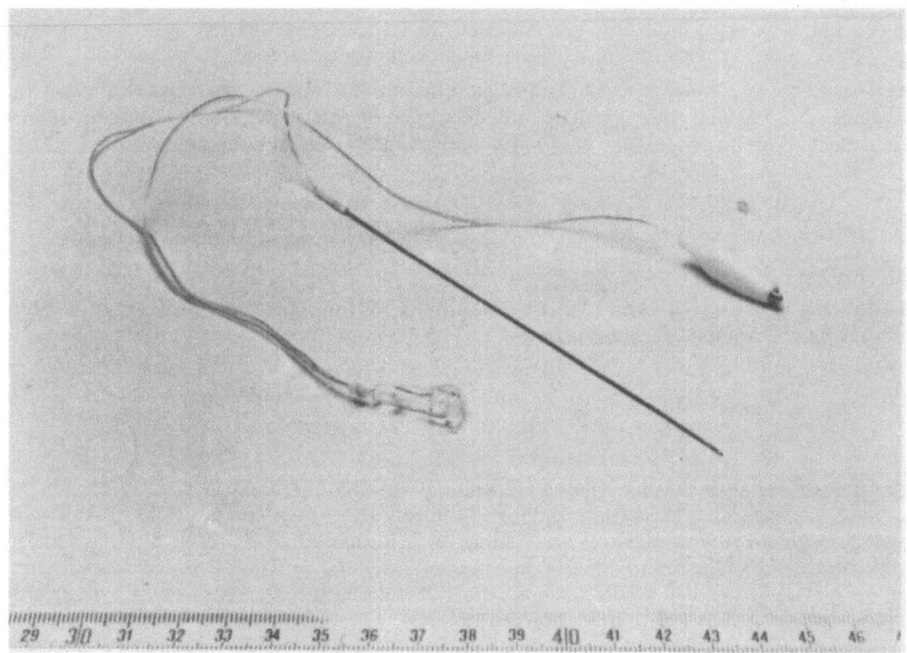

Fig. 19: Needle electrode, insulated exept for the tip for use in direct stimulation of nerves (scale in cm). From JENKNER [118].

II. Special Section

Application of electric impulses of the specified kind (page 21) to influence sympathetic nerves may be understood easily. It is possible to interrupt the function of a sympathetic nerve by applying electrodes in a specific way, specific as to size and position of the electrodes. To ascertain if a proposed measure will be followed by the desired effect, one should test the respective nerve block by monitoring the circulation before prescribing a series of therapeutic treatments. As an example, the case of a patient who presented leg pain about 1 1/2 to 2 years after a disc operation is cited. The pain was the same as it was before his operation. Checking the identical leg for either a recurrent disc protrusion or a so-called post-sciatic circulatory disturbance, the hemodynamics of the leg were examined, using an appropriate method (rheogram). This approach is the simplest in arriving at a clear diagnosis. For a rheographic [136] study an AC-field is used and the impedance changes registered (fig. 20). After 10 minutes of exposing the lumbar sympathetic chain of the respective side to the named electric impulses, the study is repeated and results are compared with the initial control, either visually (fig. 20) or, for a more exact and objective evaluation, with computer assisted evaluation (fig. 21). In the case mentioned the print out of the test result registered an optimal return to normal circulatory conditions. Accordingly a treatment session of the electric block of the sympathetic chain at the L-3 level (as in the test) was prescribed. After a month of daily 20 minute treatments, the patient was pain free and remained so after one year, at which time the test was repeated. If this test had presented a negative result, other investigations would have been made to obtain a clear diagnosis. The symptoms dictate which segment has to be used for testing. In the afore mentioned case, electrode positions were chosen which may be recognized from figs. 29 - 32. Such tests may be performed at all parts of the sympathetic system even if circulatory symtoms are not the primary complaint. In our case it was pain. It may well be tinnitus, vertigo, swelling or any other symptom related to sympathetic dysfunction. Circulatory studies are necessary for testing because they allow objective statements. Where to place electrodes in what condition will be described on the following pages.

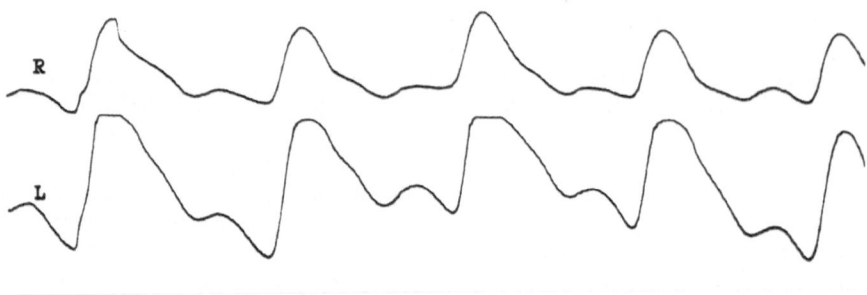

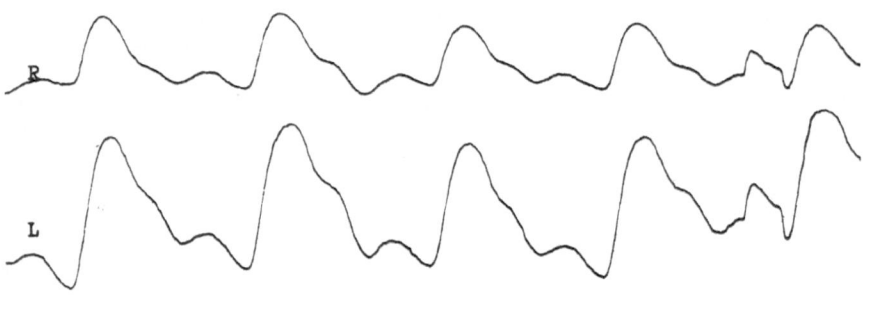

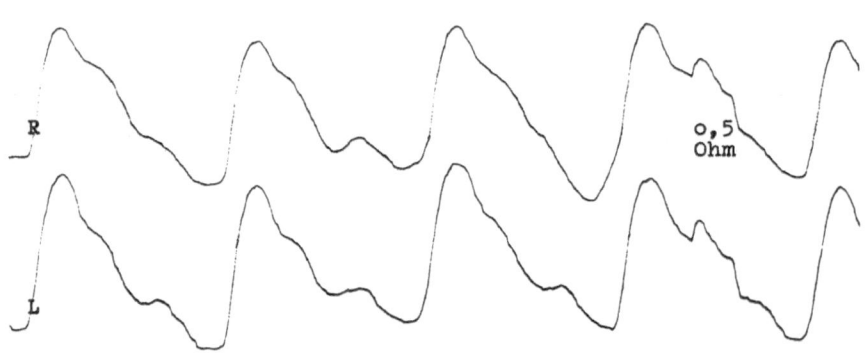

o,5
Ohm

62

Segmental Access: All sympathetic fibers to this ganglion are derived from the 1st thoracic segment via the stellate ganglion.

Anatomy: By its position along the longus capitis muscle ventral of the transverse process of the 2nd and 3rd (at times also 4th) cervical vertebrae, it is most suitable for electric blocking due to its closeness to the skin. It is situated close to the loose fibrous tissue surrounding the internal carotid artery and the jugular vein. Close to the cephalad part of the ganglion, the hypoglossal and glossopharyngeal nerves and the laryngeal and pharyngeal branches of the vagus nerve are found. The ganglion is supposed to have resulted from unification of the four uppermost ganglia of the cervical sympathetic chain. Fibers from this ganglion are passing via the following nerves to the named organs: The grey communicating branches reach the 1st, 2nd, 3rd and 4th somatic cervical nerves; sympathetic roots to the hypoglossal, vagus (jugular ganglion and nodose ganglion; the latter with many small and large branches) and glossopharyngeal (petrose ganglion) nerves; ciliary ganglion, laryngopharyngeal plexus, esophagus, superior cardiac nerve, phrenic nerve, superior laryngeal nerve (before it divides into an internal and external branch), internal carotid nerve to the internal carotid artery, where the internal carotid plexus is formed, from which the plexus to the superficial temporal artery is derived; plexus of the facial artery; to hypophysis, deep petrous nerve, cerebral vessels, vessels of eye, parotid gland, submaxillary and submandibular glands; all peripheral branches of the external carotid artery, cranial parathyroid nerve and finally the large connecting bridge to the middle cervical ganglion via the sympathetic chain. Also to the otic ganglion, the bodies of the second and third cervical vertebra and the muscles of the neck with a posterior branch for each (fig. 22a and c).

Positions and Sizes of Electrodes: Place the anode (1,7 cm^2) 1 cm caudal and 1 cm anterior of the mastoid process; and the cathode on the opposite side of the body 3 cm caudal of the angle of the lower jaw. The cathode should be 7,3 cm^2 in size. For the block, the patient should either be in a reclining position with the head supported by a small pillow and turning the head towards the opposite side or in a sitting position also with the head turned towards the other side (fig. 22b). For natural sizes of electrodes see appendix. Position of electrodes is shown in figs. 23 and 24.

Fig. 20: Longitudinal rheogram of both legs [120] for testing effect of a lumbar sympathetic block in a patient suffering from post- sciatic circulatory disturbances of right leg (late sequelae of a herniated disc two years prior to this with pain also in right leg). From top to bottom (in each group of recordings) sequence of tracings is: EKG; R = rheogram of right leg; L = same of left leg. - Uppermost group represents pre-testing control; center group was recorded just after current was switched on, as shown by artefacts in EKG lead; such artifacts have also been described by HAUPTMANN, [88]; lowest group of tracings was recorded 10 minutes after beginning of current flow. Note equal amplitudes now as compared to control situation. See also fig. 21. From JENKNER [110].

Fig. 21: (overleaf)Computer evaluation of tracings of test in fig. 20. Note improvement of right relative pulse wave volume by 78% and angle of inclination (index N) by 52%. From JENKNER [110].

```
NEUROSURGERY CLINIC
AMBULATORIUM SUED
VIENNA,AUSTRIA

PREVIOUS RECORD:
RUNNING NUMBER AND AGE:      ?213841
SEX AND KIND OF RECORD:      ?MB1
ON DISCETTE:?DY1
THIS RECORD:
RUNNING NUMBER AND AGE:      ?213941
SEX AND KIND OF RECORD:      ?MB1

CONCERNS:post-sciatic              DATE  7-DEC-1983
circulatory disturbance ,
rt leg in a male 45 yrs
2 yrs after lumbar disc
surgery


                                   TO DR.



DEAR COLLEAGUE:


THANK YOU FOR REFERRING THE ABOVE NAMED PATIENT FOR
A RHEOGRAPHIC STUDY OF

             RHEOGRAM OF BOTH LEGS

WHICH WAS PERFORMED TO-DAY.
THE TRACINGS RECORDED AT REST REVEALED THE FOLLOWING :


PREVIOUS RECORD:

             a    b    c    d    e    A    B    C    K    Pi    N
     RIGHT   27   23   45   14   19  120   10   21   28   62    81
     LEFT    34   23   44   11   22  202   46   55   25  119   134

THIS RECORD:

             a    b    c    d    e    A    B    C    K    Pi    N
     RIGHT   30   23   45    7   25  213   55   58   25  129   123
     LEFT    33   23   43   11   23  200   41   50   24  121   132


CONTROLLING THE RECORD AFTER
        1o min. of electric lumbar sympath.block,rt.

REVEALE A CHANGE OF

     REL.PULSE VOLUME    PLANE INTEGRAL                 N
     RIGHT    78 %           106 %                      52 %
     LEFT     -1 %            2 %                       -1 %

CONCLUSION : Most effective therapy. Suggestion:such electric block
should be carried out daily for 2 weeks. Complete recovery is to be
expected.

                         WITH KIND REGARDS



                         F.L.JENKNER,M.D.
                         F.I.C.S.,F.N.Y.C.S.
                         PROF.OF NEUROSURGERY
```

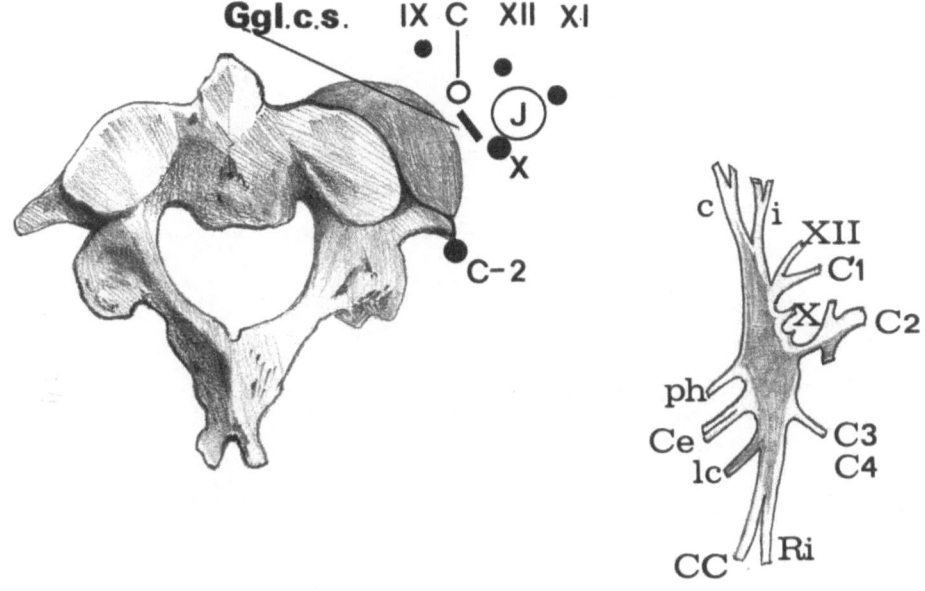

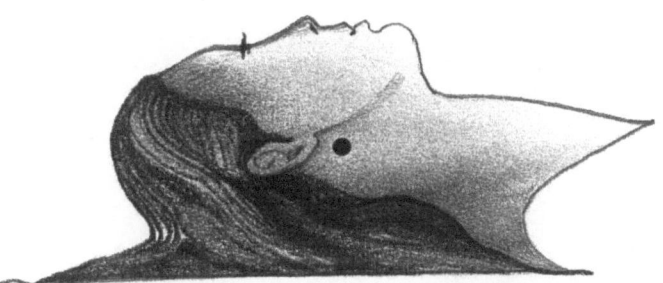

Fig. 22:   (a) Anatomic sketch of superior cervical ganglion (Ggl.c.s.) and its surrounding structures which also may be influenced by a current. Symbols indicate: IX, X, XI and XII cranial nerves; C = internal carotid artery; J = internal jugular vein; C - 2 = second cervical nerve. The second cervical vertebra was drawn as seen from above and the other structures in their position they occupy in a transverse section of the neck at a level of the second cervical vertebra. (b) Marking the position of an anode for blocking this ganglion. (c) Longitudinal section through the ganglion to illustrate connections to other structures the position of which was shown in (a). Symbols mean: c = internal carotid nerve; i = jugular nerve to extracranial ganglionn of glossopharyngeal nerve and jugular ganglion of vagus nerve; XII = accessory nerve; X = connection to nodose ganglion of vagus nerve; C1, C2, C3 and C4 = communicating branches to ventral branches of cervical nerves 1 - 4: Ri = caudal interganglionary branch; CC = cranial cardiac nerve; lc = branch to cranial laryngeal nerve; Ce = external carotid nerve; ph = pharyngic branch. (c) after RAUBER - KOPSCH [l.c. 110].

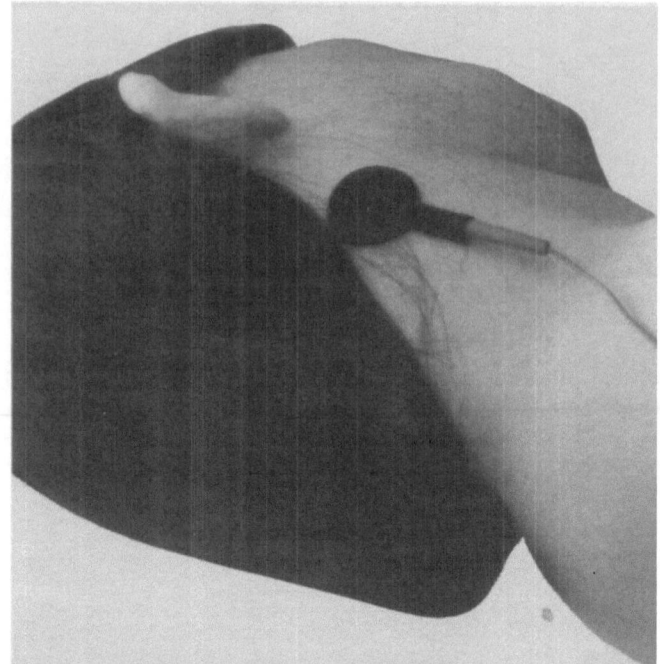

Fig. 24: Position of cathode to block the opposite superior cervical ganglion; meaning, for blocking the left ganglion, the cathode has to be placed on the right side.

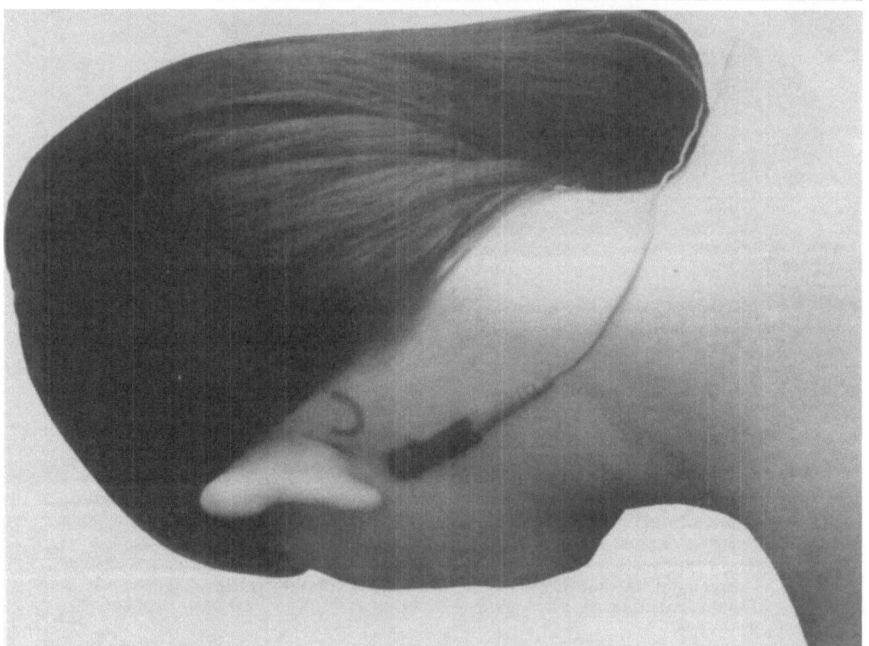

Fig. 23: Position of anode for blocking the superior cervical ganglion on left. Contour of mastoid process is marked on skin.

In one third of all cases, the effect of this block should already be visible after the first therapeutic session, while in another third, this should occur after the second block; the remaining third of the patients submitted to this electric block will present the HORNER syndrome after the third session. It is therefore necessary to observe the patient's pupils. A HORNER syndrome consists of miosis, ptosis and enophthalmus, often accompanied by conjunctival injection, reddening of the homolateral side of the face with anhidrosis of face, arm and hand ipsilaterally; increased lacrimation, increased skin temperature of face, arm and hand and nasal congestion. As a side effect of this block, the fibers of the vagus nerve may also be blocked concomittantly, resulting in increased pulse frequency; this increase may at times be coupled with a subjective sensation of tachycardia. If this is the case, the block should be terminated and continued as a stellate block since this side effect does not occur with a stellate block. See also page 69.

Indications: This block is executed only rarely and then mainly by otolaryngologists since all its effects may be achieved by blocking the stellate ganglion. In spite of this, the block is mentioned here as an additional and interesting possibility. The block is successfully executed for vasospastic states of the head, headache and migraine. Cerebral and meningeal vessels are dilated thereby. A sensation of a lump in the throat due to vascular spasms will be relieved, but not so if caused by psychogenic factors. Circulatory disturbances of the inner ear, concerning the finest branchings of the cochelar as well as vestibular nerves are a well known indication just as vertigo and tinnitus; for the last two, however, we advise to treat by stellate blocks, with simultaneous i.v. infusions (see page 73). We advise to check the efficacy of this block on cerebral hemodynamics, preferably by a method free of influence on circulation such as rheoencephalography [110]. If, after a 10 minute exposure to the electric block, an effect, however slight, is visible, then the block represents an effective means of therapy; if this is not the case, the diagnostic approach must be revised and reconsideration of the problem becomes necessary.

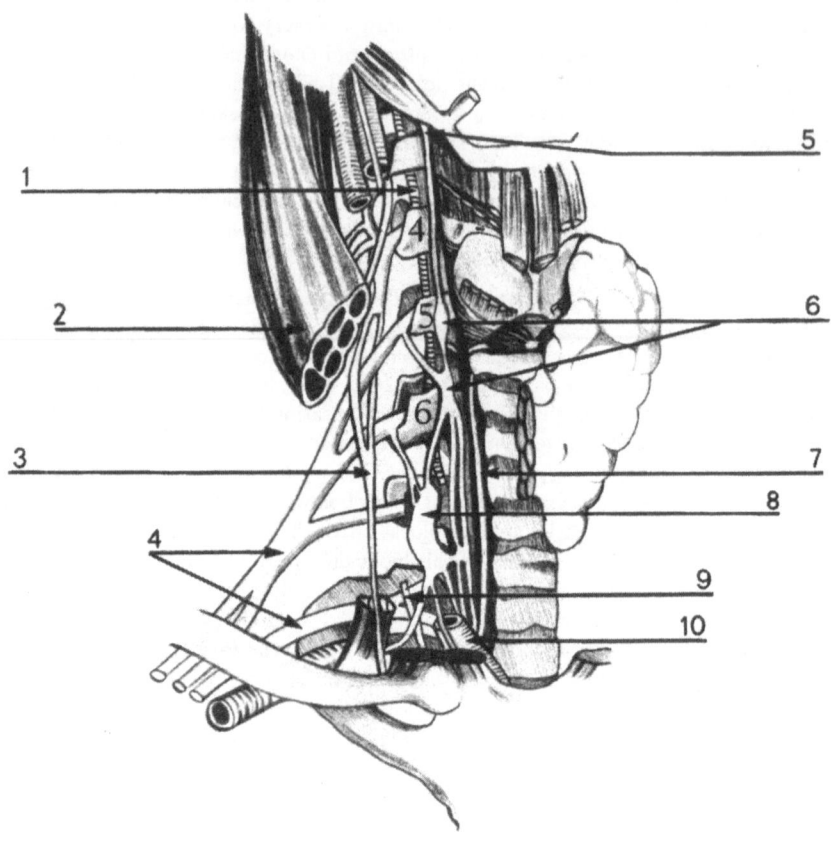

Fig. 25: Anatomic sketch to show position of stellate ganglion (rt). 1 = vertebral artery; 2 = sternocleidomastoid muscle; 3 = phrenic nerve; 4 = brachial plexus; 5 = superior cervical ganglion; 6 = middle cervical ganglion; 7 = recurrent nerve; 8 = stellate ganglion; 9 = vagus nerve and 10 = common carotid artery. From JENKNER [118].

STELLATE GANGLION

Segmental Access: Fibers for this ganglion supplying the regions of head, neck, shoulder, arm and partially thorax are derived from the first thoracic ganglion of the sympathetic chain.

Anatomy: This ganglion is an oblong structure situated in front of the transverse process of the seventh cervical vertebra and the head of the first rib. Its position very close to the skin makes it very apt for electric blocking. Close to it, one finds the recurrent nerve and parts of the brachial plexus (fig. 25).

Positions and Sizes of Electrodes: Place the anode (1,7 cm²) where the needle would pierce the skin according to HERGET, which is just anterior to the anterior border of the sternocleidomastoid muscle. The cathode (96,8 cm²) should be fixed in the midline on the back over the dorsal spinous processes of the lowest cervical spine and the upper three thoracic vertebrae. The longitudinal axis of it should be the mid-sagittal line. This position is correct for either side and for bilateral application which may be done (contrary to a needle block). Especially for home treatment the use of karaya is advised for ease of application without an assistant and because no cleansing of skin needs to be done after use.

Judging the efficacy of the block is done by paying attention to the pupils, just as was described for the superior cervical ganglion block. A HORNER'S syndrome should become apparent. As side effects, only concomitant blocking of the recurrent nerve (because of the anatomic proximity; see paragraph on ANATOMY, last sentence) with its resulting coarse voice is seen. There is no effect on blood pressure - contrary to a needle block. We are unable to give a reason for this phenomenon. In spite of rigorous controls we could never observe a fall in blood pressure. Therefore, patients may drive a car immediately after this block, which after a needle block we did not allow for an interval of 24 hours.

Indications: The pathophysiology of sympathetic overactivity points to the many and rewarding applications of blocking this ganglion electrically; for many years, the author does not use needles any longer, except in case of the only contraindication (see pg. 56) i.e. in patients wearing a demand pace maker. All vasospastic states existing in the area of supply of the ganglion may be influenced by this block.Such conditions are e.g. cardiac diseases, especially those showing a depression of the ST-portion of the ECG. This part of the ECG is elevated by a stellate block back to normal level. Also in cardiac decompensation circulation of coronary arteries will improve. In addition, it has been proven [173] that in cases of acute infarction the cardiac performance will be increased by about 3o%, lactate metabolism is normalized; paroxysmal tachycardias may be diminished if not interrupted by a stellate block. One group of investigators [16] advises the use of this block preoperatively before surgical procedures on the open heart to prevent postoperative increase of blood pressure. According to KAADA et al. [135] it is indicated also in cases of slight and moderate hypertension to decrease blood pressure.

One may advise this block also for some forms of asthma to prognosticate the possible effect of a cervico-thoracic sympathectomy and differentiate some causes of a cervical syndrome. A positive influence results for vasospastic states of head, face, arm, lung; arterial dysfunction (Raynaud's and Burger-Winniwarter's diseases, Volckmann's contracture and as an adjuvans in scalenus anticus syndrome); thromboses and embolisms to arm, lungs, brain and venous dysfunction (such as lymphedema of the arm as it so frequently occurs after surgery of the female breast; early blocking will have much better results) of a thrombotic type or postphlebitic edema and mixed forms; cold injury of arms, nose, ear; Meniere's syndrome and disease; cervical migraine (but not if there exists a malposition of the cervical spine which ought to be looked for and would need manual adjustment [119, 121]).Most essential is the early blocking of the stellate ganglion in cases of cerebral microembolisms where density changes in CT will disappear.

The improvement of values for peripheral vascular resistance in rheoencephalographic studies (see table 7) and normalisation of "wash-out" in Tc-scans (see fig. 26) has been published [7] and is in agreement with the opinion held by well-known neurologists [6].

peripheral vascular resistance in dyn/sec/cm^{-5}			
	before	after	change	remark
3 cases of non-responders	46	51	+11%	worse
7 cases of responders	57	33	-42%	much better
hemispheric flow in ml/min			
	before	after	change	remark
3 cases of non-responders	151	133	-12%	worse
7 cases of responders	204	300	+47%	much better

Table 7: Peripheral-vascular resistance and hemispheric flow in 10 cases of thalamic pain (7 cases of responders and 3 cases of very severe changes) due to circulatory disturbances in basal ganglia. Control values and change after 4 weeks of electric stellate block on appropriate side. From ATEFIE and JENKNER [7].

It has been shown objectively [115] that a stellate block will close all arterio-venous shunts in the area of supply of the respective part of the sympathetic system, meaning also within the brain. Therefore, most convincing effects are seen in all those instances where this mechanism of open a.-v. anastomoses is existing. One of the most frequently seen conditions of this kind is "migraine" [HEYCK, 94]. If the state called "migraine" by a patient in effect really is caused by this mechanism – as postulated by HEYCK – we advise daily treatments by electric stellate blocks for 2o minutes each through two months and see that two additional measures are executed: (1) Fenfluramine hydrochloride 2o mg before going to bed should be prescribed for 6 weeks (i.e. one package of 5o tablets) the reason being that this substance empties all serotonine stores of the body (as of GILMAN AND GOODMAN). And (2) all women taking contraceptives should be advised that they should use a type of hormonal preparation neutralizing or equalizing their hormonal metabolism (or type). We are well aware that gynaecologists are negating this influence [15]. But as for headaches of this type discussed here, this important aspect of the hormonal type of a woman may not be negated. Accordingly, females, the physique of which signifies

70

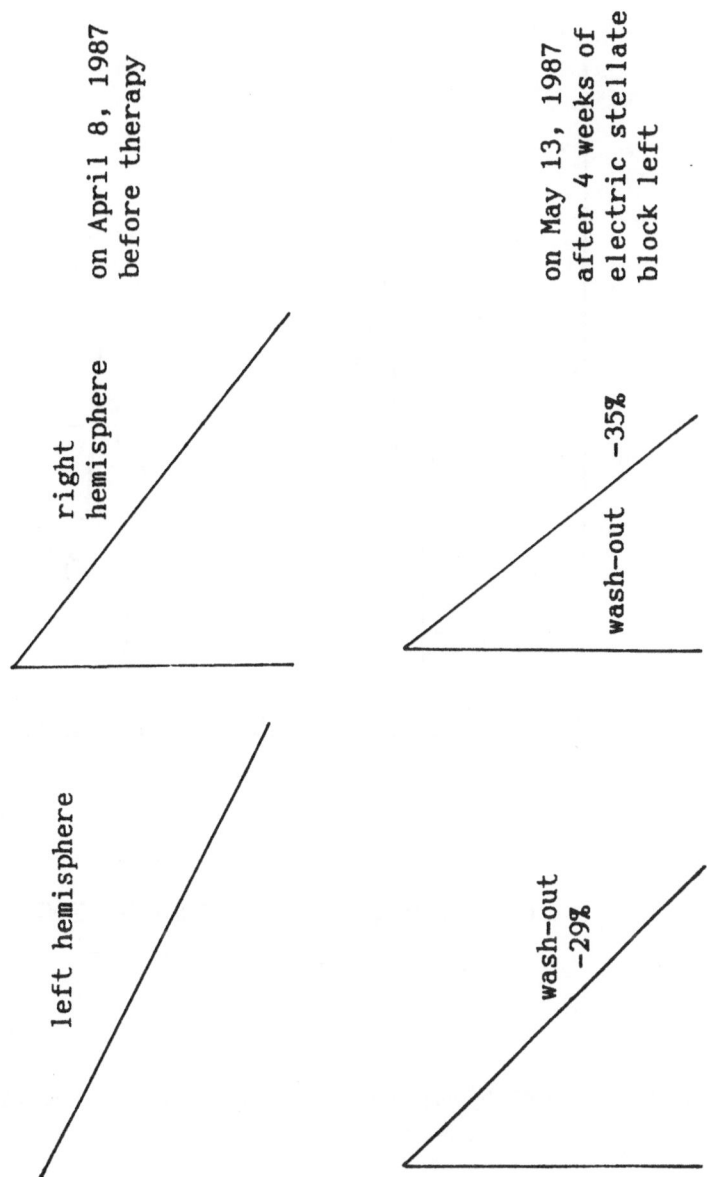

Fig. 26: Tc_{99m}-brain scan before (left; markedly prolonged "wash-out" period) and after (right) a 20 minute electric stellate block home therapy every day/1 month. Note diminished time of wash-out period (by 35%) meaning approaching normalcy. From ATEFIE and JENKNER [7].

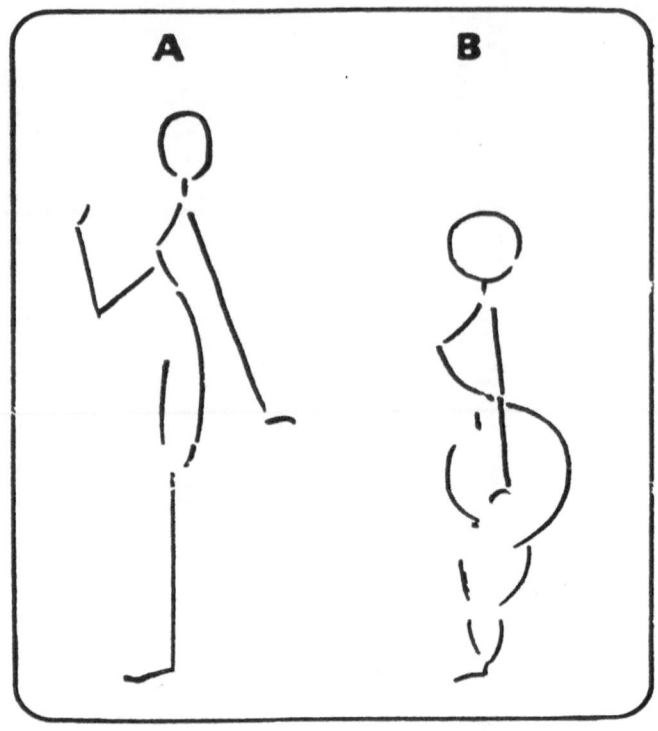

GESTAGENE ESTROGENE

TYPE TYPE

Fig. 27:   Schematic drawings of two most common types of female body shape.

gestagene accentuation (fig. 27) and who are suffering from migraine much more frequently should be prescribed a contraceptive of accentuated estrogene content while women with an estrogene accentuated physique should rely on gestagene accentuated contraceptives. If this equalizing of a hormonal type does not exist (with the medication the patient is using) we change the medication accordingly. This measure by itself will reduce frequency of attacks of migraine.

However, this applies only if migraine is caused by the named mechanism of open anastomoses. If the cause of migraine is to be found in malposition of the cervical spine between the segments C-5 and C-6 (reason to stress this segmental level: in the inter-vertebral foramen C5/6 there is situated the vertebral ganglion, from which arises the vertebral nerve = the sympathetic fibers accompanying the vertebral artery), readjusting the position to a normal lordosis will be treatment of choice [119].

Other indications where this block proved very valuable are: posttraumatic bone atrophy (SUDECK [127], fig. 33) just as are all other cases of sympathetic dystrophies, e.g. posttraumatic osteoporosis, causalgia, phantom pain, stump pain. Hyperhidrosis of upper half of body. As an adjuvant in the therapy of all those states where interruption of sympathetic pathways will improve the condition as e.g. in ulcers showing a missing tendency of healing, post-zoster-neuralgia, anginoid states; sometimes asthmatic states or after plastic surgical procedures and all other circulatory disturbances existing in arms, neck and face. Most effective is this stellate block in the therapy of erythema pudendum. In cases of vertigo and tinnitus, we advocate a combination of this block with an infusion containing neurobion forte (vitamin B-1 1oo mg, B-6 1oo mg, B-12 1ooo mg), cocarboxy-lase (5o mg), tioctane 50 mg, bupivacaine hydrochloride (1o ml of a 1% solution) in 25o ml low molecular dextrane (Rheomacrodex) except in hypertonic patients, where we use "Solcoseryl". During the initial phase, block and infusion will be synchronously while the infusion continues to its end after the 2o minutes of blocking. As an added measure, this block has proven its value in cases of pulmonary and/or cerebral edema. We have observed several cases of quadrant hemianopia in which electric stellate block has begun immediately after the onset of visual disturbances and was carried on for a month with daily sessions: hemianopia disappeared entirely, and did not recur.

Finally we should like to point to the possibility of improving circulation of the cervical cord: In a case where disturbed circulation has been the cause of a motor (anterior column) lesion at C-6 on left (proven electrophysiologically) and where gradual decrease of motor function was occurring, daily electric stellate block could make the subjective symptoms disappear after a three month therapy period and electrophysiologically, no more evidence of a motor C-6 lesion could be demonstrated.

THORACIC AND LUMBAR SYMPATHETIC CHAIN

Segmental Access: Segmentally organized supply to or from the somatic segmental nerves from D-1 to 12 and L-1 to 5 (fig. 28).

Anatomy: The sympathetic chain is situated at the antero-lateral border of the vertebral bodies. Since the circumference and size of the vertebral bodies increase from upper thoracic to lumbar spine, the direction of field lines of the electric field, appilied from the skin, must have the same oblique direction as a needle has to have when blocking by needle. In a caudad direction, this angle of obliqueness should increase. See also under "POSITION and SIZE of ELECTRODES". One should keep in mind, that some somatic afferent fibers are crossing through sympathetic paravertebral ganglia without being interrupted by a synapse there. Therefore, these fibers are blocked also, but this is not as important as it used to be for blocking by needle, since the expanse of an electric field will always block somatic fibers simultaneously with a sympathetic thoracic or lumbar segmental nerve block.

Positions and Sizes of Electrodes: A 7,3 cm^2 anode should be placed paravertebrally from 7 to 10 cm from the midline: from a cranial to the most caudal position, this distance should increase. A 96,8 cm^2 cathode should be placed ventrally, at the highest thoracic level directly in a midline position, at the lowest lumbar L-5 level markedly lateral towards the contralateral side (compared to the anode). See figs. 29 and 30. The use of karaya as a conductive medium is advocated.

Indications: The application of a THORACIC SYMPATHETIC BLOCK is advised in all conditions which are associated with spasms and circulatory disturbances as well as increased tension of fibrous capsules of parenchymatous organs (e.g. pancreas). The respective segmental level of anodal position may be recognized from a table (page 44/45). This ascertains a correct placement for the various possibilities existing for the heart, thoracic aorta, esophagus, liver, gall bladder, stomach, pancreas, spleen, kidney, duodenum, abdominal aorta inclusive its branches, ascending and transverse colon, in part also for ureters, prostate, testes, ovaries and uterus (including for dysmenorrhoea): one recognizes a great variation of symptoms which may be treated from the thoracic chain.

Main indications for the LUMBAR SYMPATHETIC BLOCK are circulatory disturbances of lower extremities (most commonly via L-3), most caudal part of aorta with branches (at L-1 and 2), descending colon, sigmoid (left position of anode), kidney, bladder, testes, prostate, ovaries, uterus and female external genitalia. For each of these organs, the anode has to be placed at the involved side and the cathode contralaterally. If possible, before a series of treatments is done, a test should be performed to ascertain the efficacy; as an example, testing before a series of lumbar chain blocks is shown in a case of postsciatic circulatory disturbance (fig. 20 and 21). For postoperative ileus, place the anode on the left side and apply 3 times daily for 20 minutes each session. Further indications are: All sympathetic dystrophies of lower extremities, (e.g. SUDECK), and lower half of body, arterial, venous and combined a.-v. diseases, such as Buerger-Winniwarter, Raynaud, other vasospastic states after arterioplasties on legs, postphlebitic edemas or edema following tumorous secondary to lesions in groins often occur after testicular sarcomata, testicular teratoma, or uterine malignancy. For these cases, we advise to keep the involved leg raised during and for 1/2 hour following a therapeutic session. Other applications are for phantom pain, if the loss of extremity was following a circulatory lesion; causalgias; as a

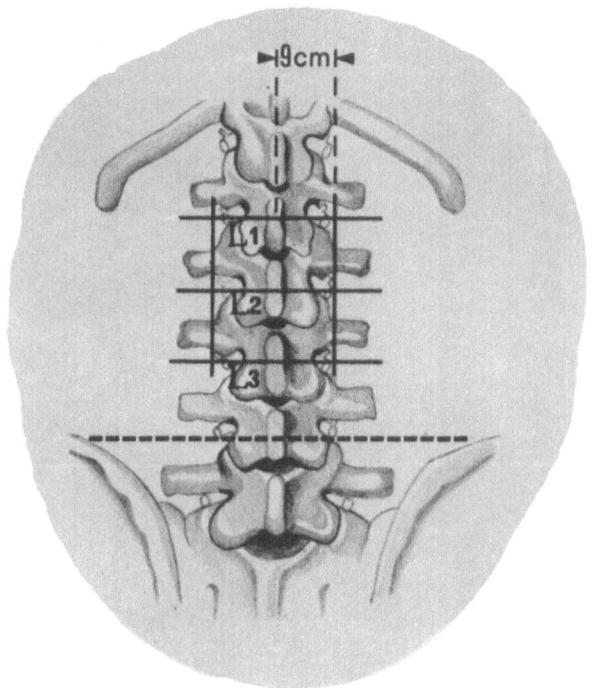

Fig. 28:   Anatomic sketch to simplify placement of anodes for lumbar sympathetic electric blocks. From JENKNER [120].

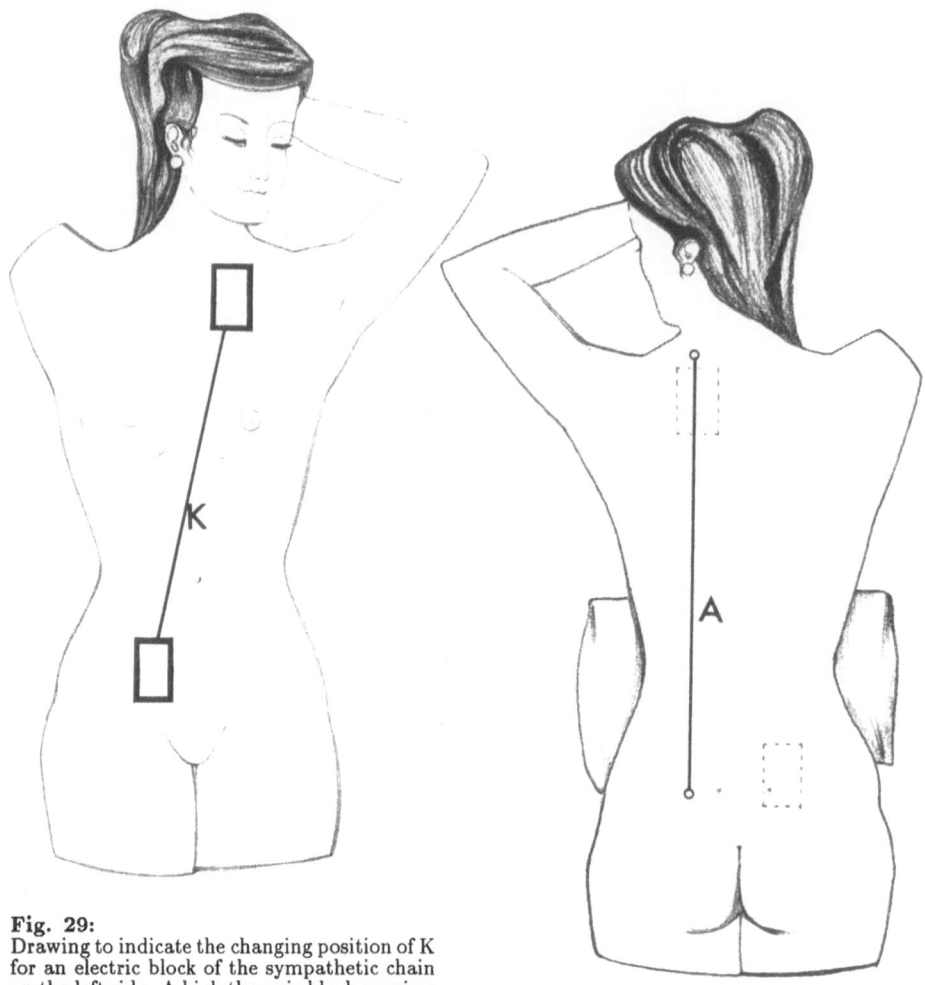

Fig. 29:
Drawing to indicate the changing position of K for an electric block of the sympathetic chain on the left side. A high thoracic block requires the cathode (K) to be placed homolaterally to the site of the anode (which is left here), for a lumbar block a contralateral cathode is mandatory. For intermediate levels select a position on the line drawn between the two rectangles.

Fig. 30:
Changing position of anode (A) from high thoracic to low lumbar (always left side) level for a block of the left sympathetic chain. Ventral positions of large cathode shown as dashed rectangles.

supplementary measure, it may be performed in cases of chronic infectious processes, badly healing ulcers, not adequate healing processes in cases of plastic surgical procedures, in stiffness of joints, hyperhidrosis of lower half of body; Hirschsprung's disease, megacolon, irritable colon, subacute and chronic pancreatitis (in case of failing effect of a thoracic block) arthritic pain, muscle spasms and cold injuries.

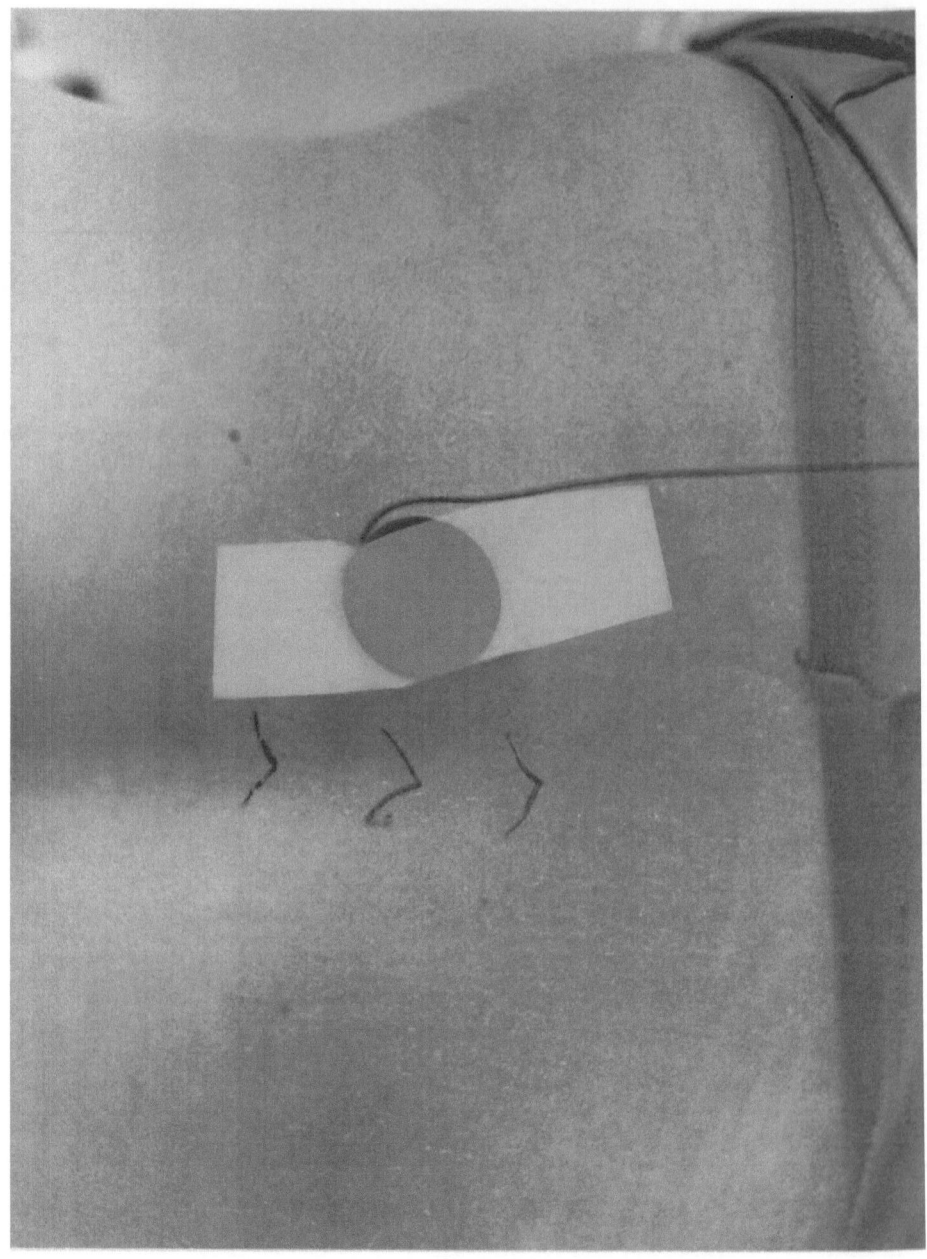

Fig. 31:   Position of anode for lumbar sympathetic block at L-3 right[120].

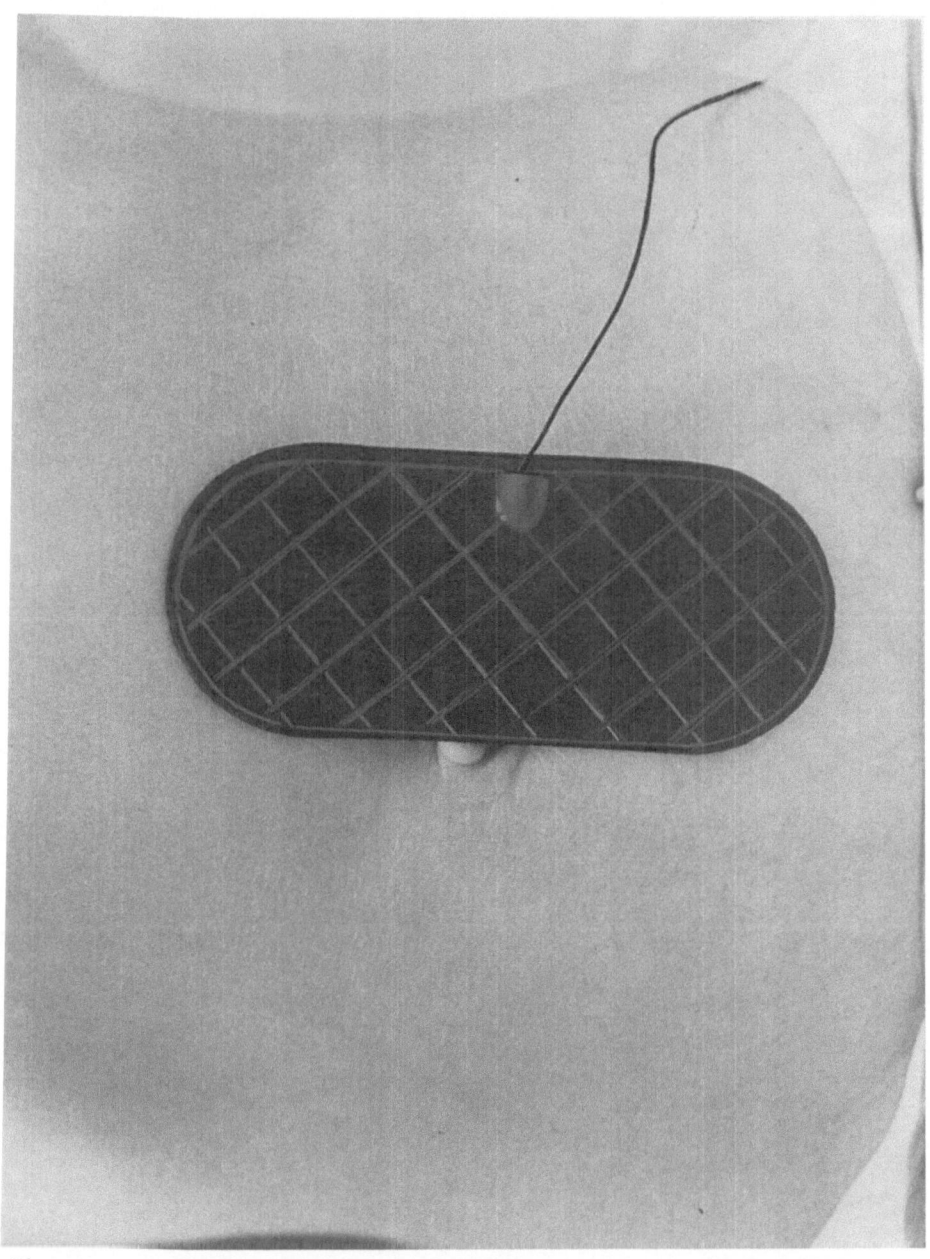

Fig. 32: Position of cathode for lumbar sympatheticic block at L-3 right, for which the cathode has to be placed on opposite side (just left of midine) with longitudinal axis in a parasagittal direction. From JENKNER [120].

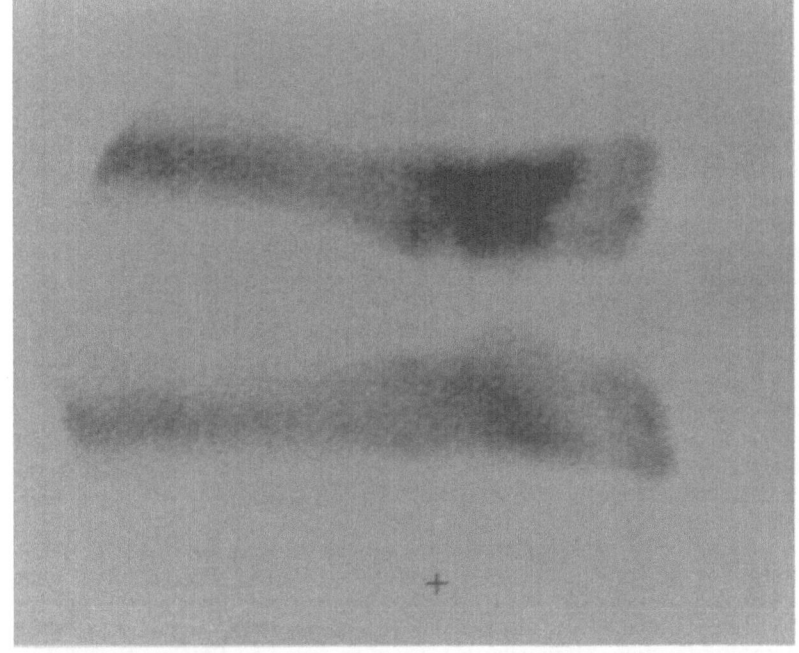

Fig. 33: Bone scan in a patient suffering from Sudeck's atrophy following a calcaneal fracture. Compare results of study before begin of therapy (left pair of films) with films obtained after three months of daily electric lumbar sympatic block (home therapy). Note that the difference in activity existing in pre-treatment condition decreased markedly (right pair of films), even though some difference remained in spite of the patient being entirely free of symptoms. Shown are static films. From Jenkner and Atefie [127].

SOMATIC NERVES

Somatic afferent nerves may not be observed as objectively for their ability to conduct (e.g. pain) as was reported for sympathetic fibers in the previous section. For these fibers, the experimentally proven effects have been substantiated by clinical reports (fig.10, 19). For somatic nerves inferences are possible only by analogy. This is made more probable by the comparison of some electric properties of somatic and sympathetic nerves (see table 1). Exposing a somatic nerve to an electric field initiated by very specific impulses via electrodes (all specified and objectivated for their effect on sympathetic fibers : see page 21) is therefore expected to have an identical blocking effect on pain conduction. The clinical observations carried out under rigidly controlled circumstances resulted in degrees of pain reduction which could be explained only if our assumption were correct. Of course a scientific proof does not seem possible at this time. This should be recalled when reading the advice given for the specific nerves in the following section.

It should be mentioned that thin somatic afferents conducting pain may (rather should) best be influenced by 35 impulses per second. It is however known that the motor fibers having larger diameters will respond better to a higher frequency of impulses. For influencing these, polarity has to be reversed, meaning that immediately over the motor nerve, a small electrode should be placed as the cathode and impulse frequency of 100 per second be used to obtain optimal effects on return of function. This should be done e.g. in treating Bells paralysis (figs. 100 - 102) or in partial loss of function of a phrenic nerve (figs. 50 - 52). A small part of the function of a motor nerve must have remained to ascertain return of function. Optimally this should be tested by EMG or NCT before start of therapy, but it may not be easiliy possible like for the phrenic nerve. In case of total loss of function return of function may not be expected by any method.

Instead of presenting pain according to topographical body areas, arrangement by nerves was decided upon. For finding involved body areas, the reader is referred either to the subject index or should refer to tables 4 and 5 and the figures between these two tables (on pages 44 through 52). There one may find help in an attempt to find those nerves which are anatomically responsible for the origin or conduction of pain to a certain area of the body. These nerves then ought to be looked up for proper electrode positions and effective therapy. For additional measures see pages 185 and 186.

SUBOCCIPITAL, GREATER AND LESSER OCCIPITAL, GREAT AURICULAR, and THIRD OCCIPITAL NERVES

Segmental Access: Suboccipital nerve from C-1, greater and lesser occipital nerves from C-2, great auricular nerve from C-2 and third occipital nerve from C-3.

Anatomy: The nerve which is most important from a clinical point of view is the greater occipital nerve (number 1 of sketch of fig. 34). It appears from under the belly of the inferior oblique head muscle at its caudal border; this is approximately in a transverse plane between atlas and axis (epistropheus), about 3-4 cm laterally from the mid-sagittal plane (number 4 of sketch). It reaches the nuchal line (number 3 of sketch) between the onset of trapezius and semispinalis muscles immediately medial of occipital artery (number 2 of sketch; may be palpated very easily), where it branches. This point of branching usually is very sensitive to pressure. The minor occipital nerve runs about 2-2.5 cm lateral and somewhat more caudal from the greater nerve (number 5 of sketch). The posterior branch of the great auricular nerve (number 6 of sketch) runs very close to the posterior face of the concha. The third occipital nerve as well as the suboccipital nerve are clinically without importance.

Positions and Sizes of Electrodes: The anode is to be placed over the exit of the involved nerve. In case of the first three nerves, its size should be 7.3 cm², as is the case also for all cathodes. For the posterior ramus of the great auricular nerve, it should be 1.7 cm² in size. Cathodes should be placed on the corresponding (mirror-like) site on the opposite half of the head. In case of unilateral pain, the anode should be over the painful side. In case of bilateral pain, the anode should be changed from one side to the other (right to left) from one therapeutic session to the other (daily, 20 minutes each). Fig. 35 presents the position of electrodes for pain in the area of the greater occipital nerve. Electrodes have to be applied with wet pads between them and the skin, since hair has very high electric resistance and would interfere with good skin contact.

Karaya may not be used for the same reason. For optimum fixation of electrodes, a rubber band is very well suited which also presses the electrodes better against the skin, thereby improving electrical contact. This position of electrodes is one of the few exceptions to the rule that electrodes should be placed at opposite surfaces of the body.

Indications: Occipital neuralgia; occipital pain extending onto the border of the hair at the forehead; pain on pressure of the exit of the greater occipital nerve either at the site indicated by number 4 or more often number 3 of the sketch. For the other nerves, the respective sites would be most painful on pressure. For differentiating occipital pain caused by pathology of these nerves from other causes like e.g. cervical syndrome. Sometimes this block is very helpful in alleviating pain in trigeminal neuralgia, either by itself or in combination with blocking the trigeminal nerve (or its respective branch). If pain is extending to neck or upper collar, the segment C-3 also is involved. If the cause of this pain (unilateral) is a gross anatomical lesion of the C-2 root, like in epidural tumor, neurinoma or similar structural changes, this block will most certainly not be effective for pain relief. Also in cases, where medullary changes at the C-2 level exclusively are present (such as circulatory disturbances on demyelinating foci of dorsal columns), it will be in vain. In these last mentioned cases, one may try to treat via the sympathetic (stellate block) and expect at least some relief. These conditions, however, are extremely rare.

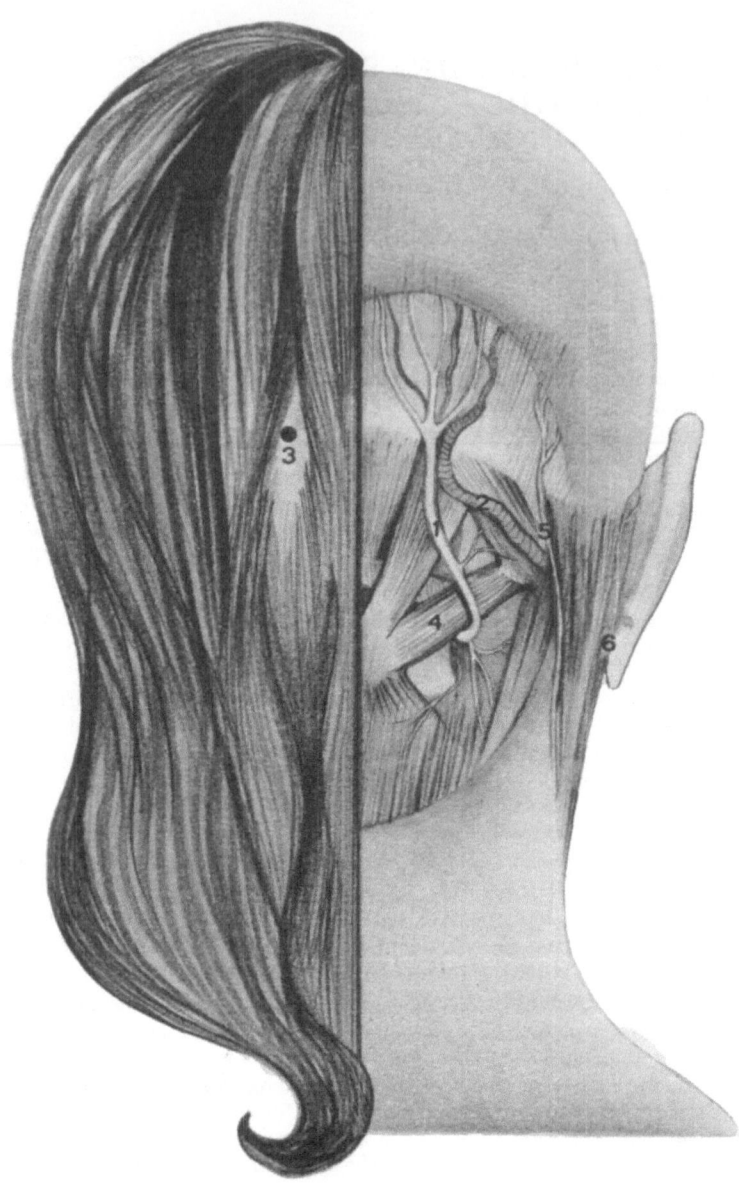

Fig. 34: Anatomic sketch on site of major occipital nerves, minor occipital and great auricular nerve. Important for placing anodes in cases of neuralgia of these nerves. Symbols are: 1 - major occipital nerve, 2 - occipital artery, 3 - site of anode placement over major occipital nerve; 4 - inferior oblique muscle of head; 5 - minor occipital nerve; 6 - great auricular nerve, posterior branch. From JENKNER [118].

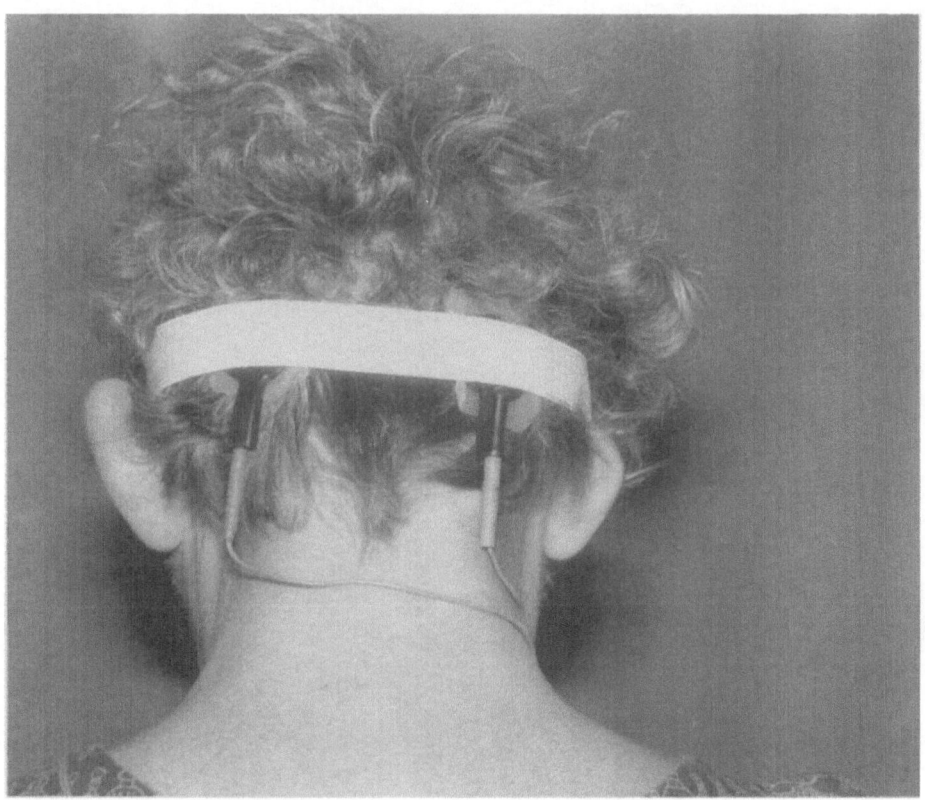

Fig. 35: Electrode position for left sided occipital neuralgia. Under electrodes wet pads are visible, to be used for better conductivity through hair by water. The left electrode should be the anode. Fixation of electrodes by rubber band.

Fig. 37: ⟹ page 85
Anatomic sketch to indicate cathode positions for the three branches of the trigeminal nerve. Full dots signify exit of respective canal the directions of which are almost contrary (arrows), but do not have any significance for electric blocks. Placement of anode of all three branches may be seen in fig. 38 to 40. From JENKNER [118].

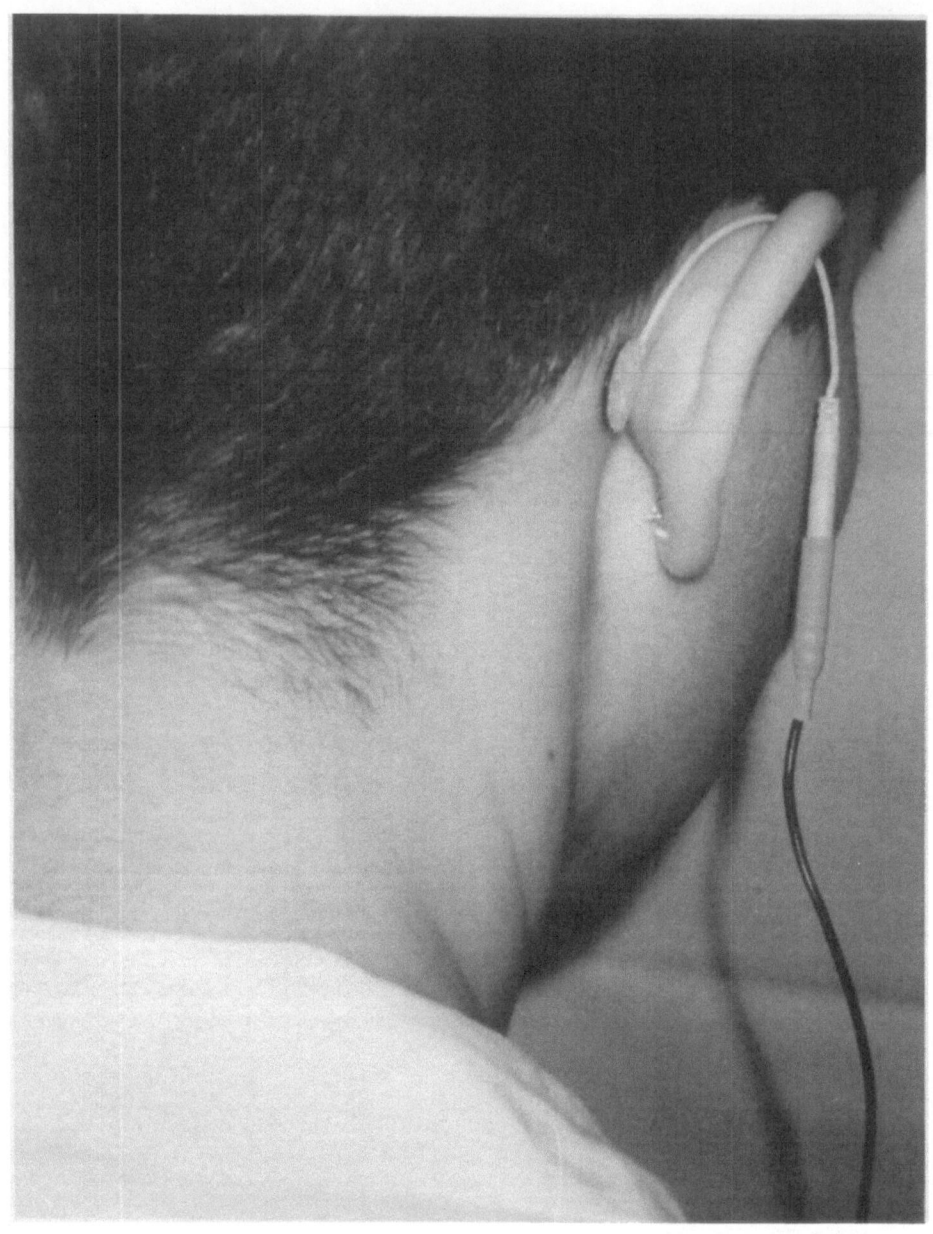

Fig. 36: Position of anode for greater auricular nerve neuralgia. Cathode to be placed on oposite side.

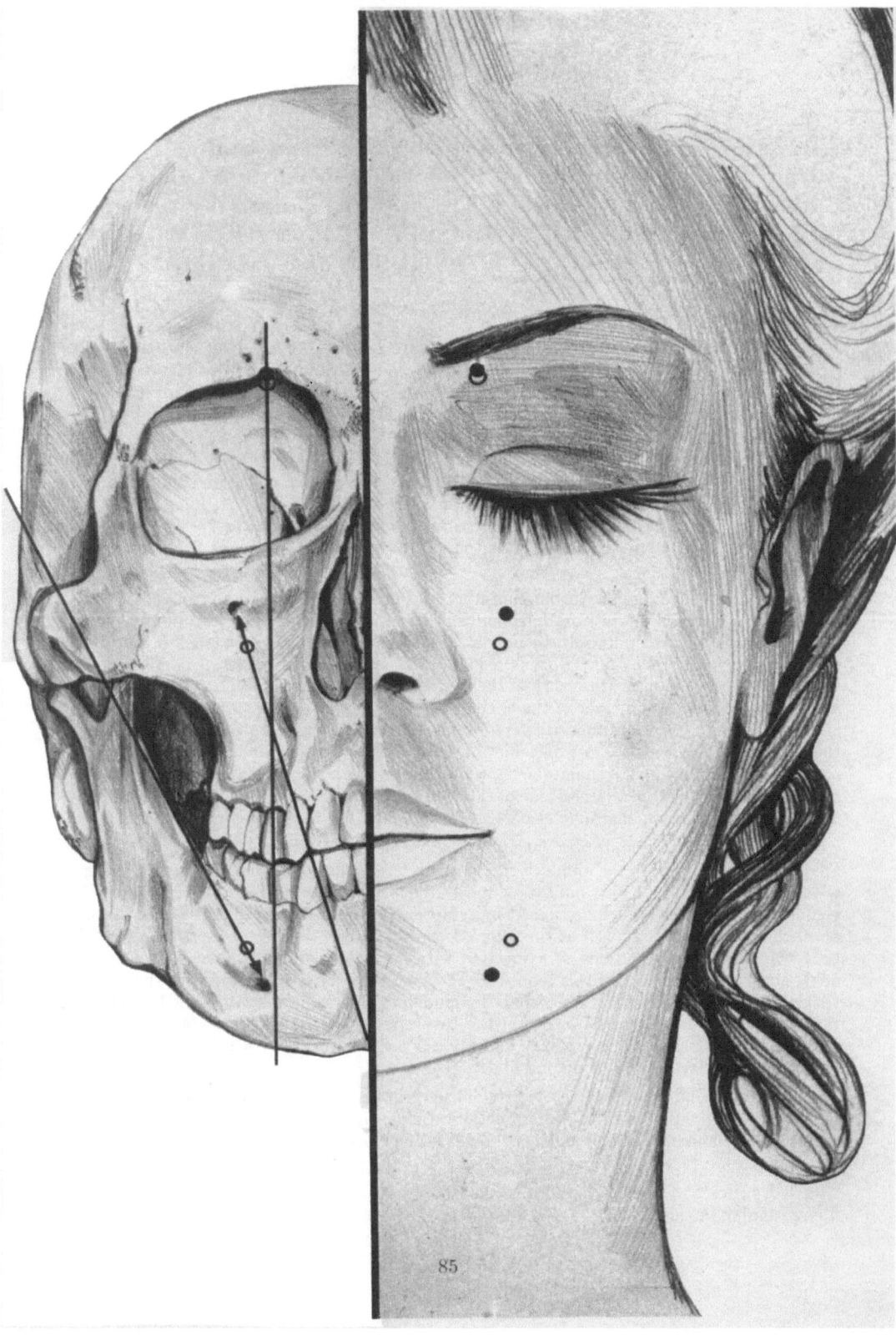

85

TRIGEMINAL (with supraorbital; maxillary and infraorbital; mandibular and mental nerves) and GLOSSOPHARYNGEAL NERVES with relatively rare NEURALGIAS OF GENICULATE GANGLION (HUNT) and SPHENOPALATINE GANGLION (SLUDER).

Segmental Access: The trigeminal nerve is responsible for carrying sensory input (touch, pain, temperature) to, and motor output (only via third division to jaw muscles) from the brain. There are two differing pain states: one, resulting from many causes, including space-occupying or malignant lesions, resulting in constant pain and loss of sensation of touch; the other, constituting genuine trigeminal neuralgia, consisting of extremely brief attacks of pain (also called "tic douloreux"), which does not include any loss of sensation of touch and no motor involvement. The course of the trigeminal tracts from cortex to mid brain nuclei and the position of these nuclei and the changing neuronal structures bordering these elements of pain conduction from the face, as well as neighboring vascular structures, allow us to understand modifications in such a way as to recognize a possible lesion.

Anatomy: The impulses from supratentorial intracranial pain receptors travel via trigeminal pathways. Those from infratentorial structures run via the glossopharyngeal, vagus and uppermost three cervical nerves. This syndrome of "end organs" has to be differentiated from other causes of such pain, such as lesions of the posterior inferior cerebellar artery or the pontine branches of the basilar artery (in the supply area of which the nucleus of the descending tract of the trigeminal nerve as well as the spinothalamic tract is situated). Such considerations are very essential for evaluating the symptoms; lesions of the descending tract lead to loss of pain- and temperature sensation homolaterally (while normal sensation of touch is maintained), while lesions of the spinothalamic tract result in sensation loss on the contralateral side of the body below the lesion. A radicular syndrome of the descending trigeminal may also suggest a lesion of the cerebello-pontine angle.

If, in addition to symptoms relating to the first branch of the trigeminal nerve, there are symptoms relating to the abducens nerve, then both symptoms originate from a lesion within the petrous pyramid, indicating most frequently the syndrome of petrositis. If symptoms of the first branch are joined by symptoms referable to the abducens, trochlear and oculomotor nerves as well as the internal carotid artery, the site of the lesion has to be within the cavernous sinus. If associated with symptoms of the abducens, trochlear, oculomotor and optic nerves, this represents the so-called orbital syndrome with the cause within the orbit. All sympathetic fibers for pupillary dilatation join the first branch of the trigeminal nerve only at the site of the cavernous sinus: only if a lesion is situated peripherally to this location, pupillary dilatation is possible together with loss of sensation on the cornea and the skin of the forehead.

For the second branch of the trigeminal nerve, there is no such combination of importance. Here, the possibility of differentiation is given as regards presence of infection of paranasal sinusses (sinusitis, mainly maxillary) or lesions of tooth or dental radices.

The third branch of the trigeminal nerve presents the possibility of pain combined with loss of force of the jaw muscles homolaterally due to the involvement of the motor fibers running in this branch.

Positions and Sizes of Electrodes: These become evident from the figures 37 to 40. For all three branches, the anode should be positioned just in front of the tragus and have a size of preferably not more than 1,7 cm^2. Cathodes have to be placed exactly over the point of exit of the respective branch through the bony skull. Their size should be about 7.3 cm^2. We advise to use karaya as a conductive substance between silicone and skin. Before acquiring a "stimulating" device, one of the important things to consider is the nature of electrode cables, which should be very flexible and as light-weight as possible; otherwise the weight of the cable may easily pull the small anode off the skin inspite of the adhesive force of the karaya.

Indications: Diffuse headaches, caused by irritation of pain sensitive nerve endings intracranially-supratentorially by various processes. Genuine neuralgia of the supra-orbital nerve is caused most frequently by herpes zoster infections of this nerve. According to STOOKEY (1959, 230) and my own [123] experience, 95 (96)% of affections of the first branch are true postherpetic neuralgia. The most frequently involved branches are the supratrochlear and supraorbital ones. Fortunately, the lacrimal and nasociliar branches are only rarely involved: this has the advantage, that the dire ocular consequences of keratitis, corneal ulcerations and panophthalmitis are seen only very seldom; the first branch of the trigeminal comprises 19% of all zoster infections (topographically) according to WULF et al. [251]. The second and third branches of the trigeminal are affected by a zoster infection in 5 (4)%. These last named branches are afflicted mainly be genuine, true trigeminal neuralgia (95%). The frequency with which the single branches are involved is the following:

branch	average age	%	male	female	right	left	pain reduct. quote (in 540 cases)	diagnosis
1	62	4	30	70	2(50)	2(50)	96 %	p-zoster
2	60	66	40	60	46(70)	20(30)	88 %	genuine
3	61	30	40	60	19(63)	11(37)	86 %	Trigeminal N.

The named quotas of pain reduction refer to mentioned diagnoses only. In cases of atypical facial pain, symptomatic trigeminal neuralgia (e.g. following tooth extraction, pulpitis, sinusitis etc.) results are significantly lower. Pain following maxillary or mandibular fractures my be treated, as may pain of malignant origin (primary or secondary blastomas of floor of mouth, tongue, jaws). In facial pain combined with pain in the neck or throat it may be combined with blocks of 3rd and/or 4th cervical nerves. Results using these combinations will decrease pain perception by about 60%. In cases of trismus one may try electric block of the third branch of the trigeminal. Only rarely, it is being advocated for differentiating atypical and other facial pain where a cause is not imminent. Blocking the relevant branch of the trigeminal nerve may be (or even should be) combined or supplemented by blocks of the greater occipital nerve in case of insufficient relief. Such a measure will not become necessary in those 10% of cases of trigeminal neuralgia which is being caused by demyelinating foci from disseminated encephalomyelitis.

Glossopharyngeal Neuralgia is much less frequently seen. But it is more often observed than the other named neuralgias (HUNT or SLUDER). It characteristically is consisting of pain in the dorsal third of the tongue (while in trigeminal neuralgia, the anterior 2/3 are involved). Also pain deep in the ears or throat are typical for it. It is not easily differentiated from HUNT's neuralgia, which in typical cases causes pain deep in the back of the nose and palate, ventral face of ear and dorsal part of ear lobe, towards the mastoid. In cases of SLUDER's neuralgia, pain is situated mainly in back of eye, over to root of nose, sometimes in and about the ear, till about 5 cm (2 in) in back of ear. It almost always is combined with nasal congestion and increased secretion. In this neuralgia, the nasociliar nerve may play a role.

Electrode Positions and Sizes: The site of placement of the anode for glossopharyngeal neuralgia is shown in fig. 41. Anode and cathode are of the size of the identical electrodes of trigeminal neuralgia. We suggest to treat cases of HUNT and SLUDER's neuralgias (if a correct diagnosis has been established) via the sympathetic by stellate blocks homolaterally.

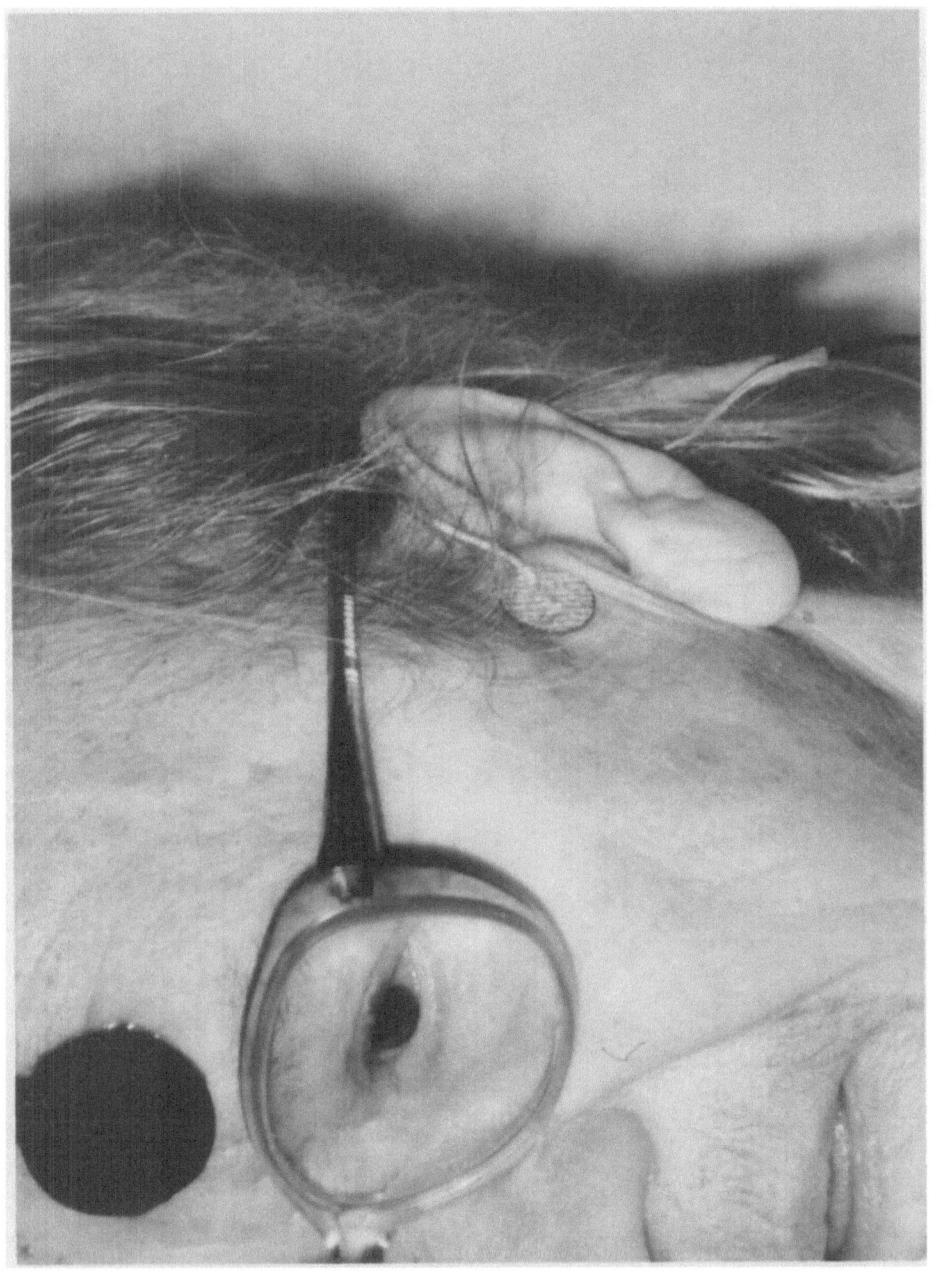

Fig. 38: Anode (black) and cathode (gray) in place for blocking the first branch of the trigeminal in a case of ophthalmic zoster.

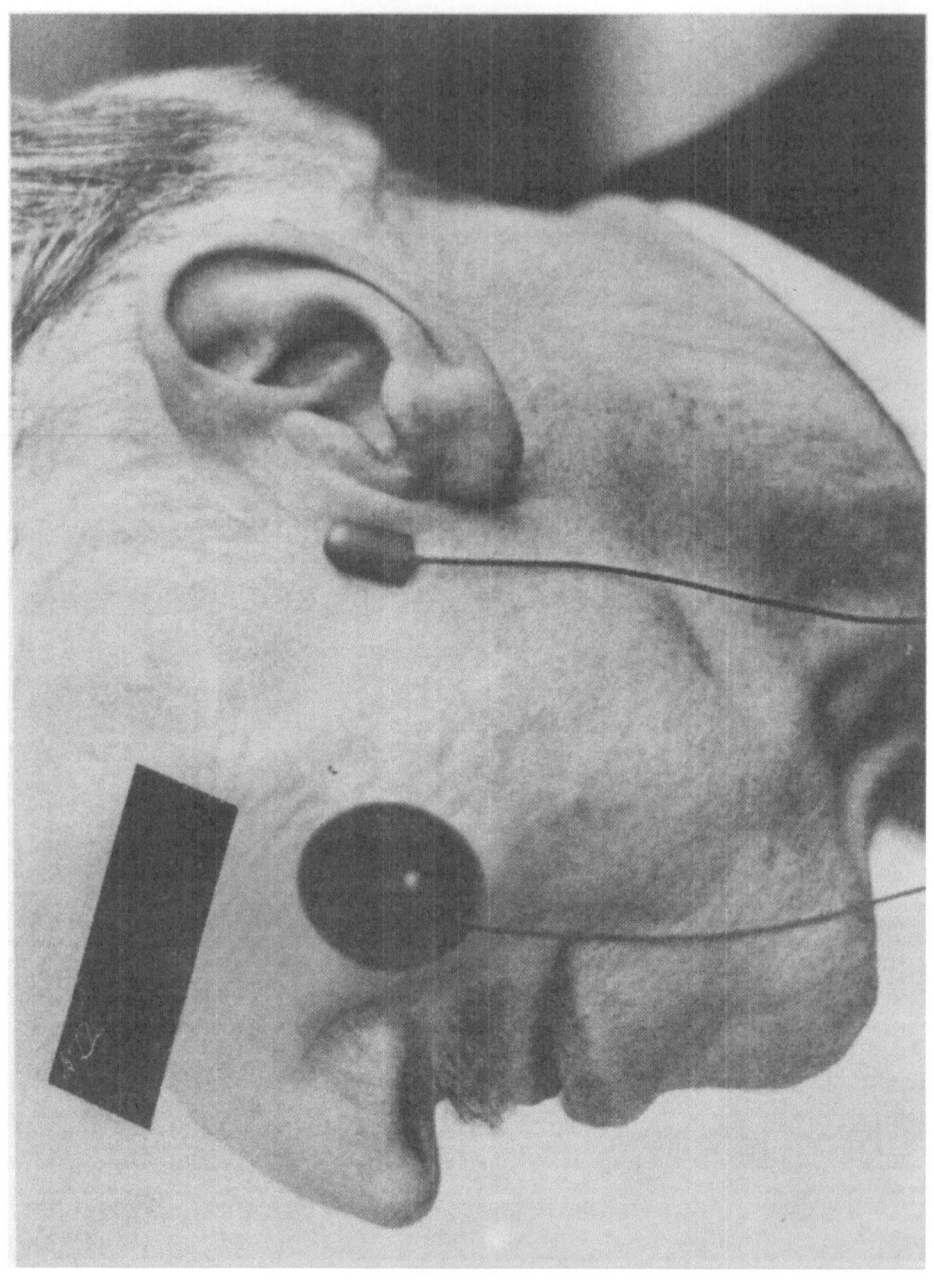

Fig. 39: Electrodes placed for tratment of trigeminal neuralgia (second branch, left side; small electrode = anode). From JENKNER [118].

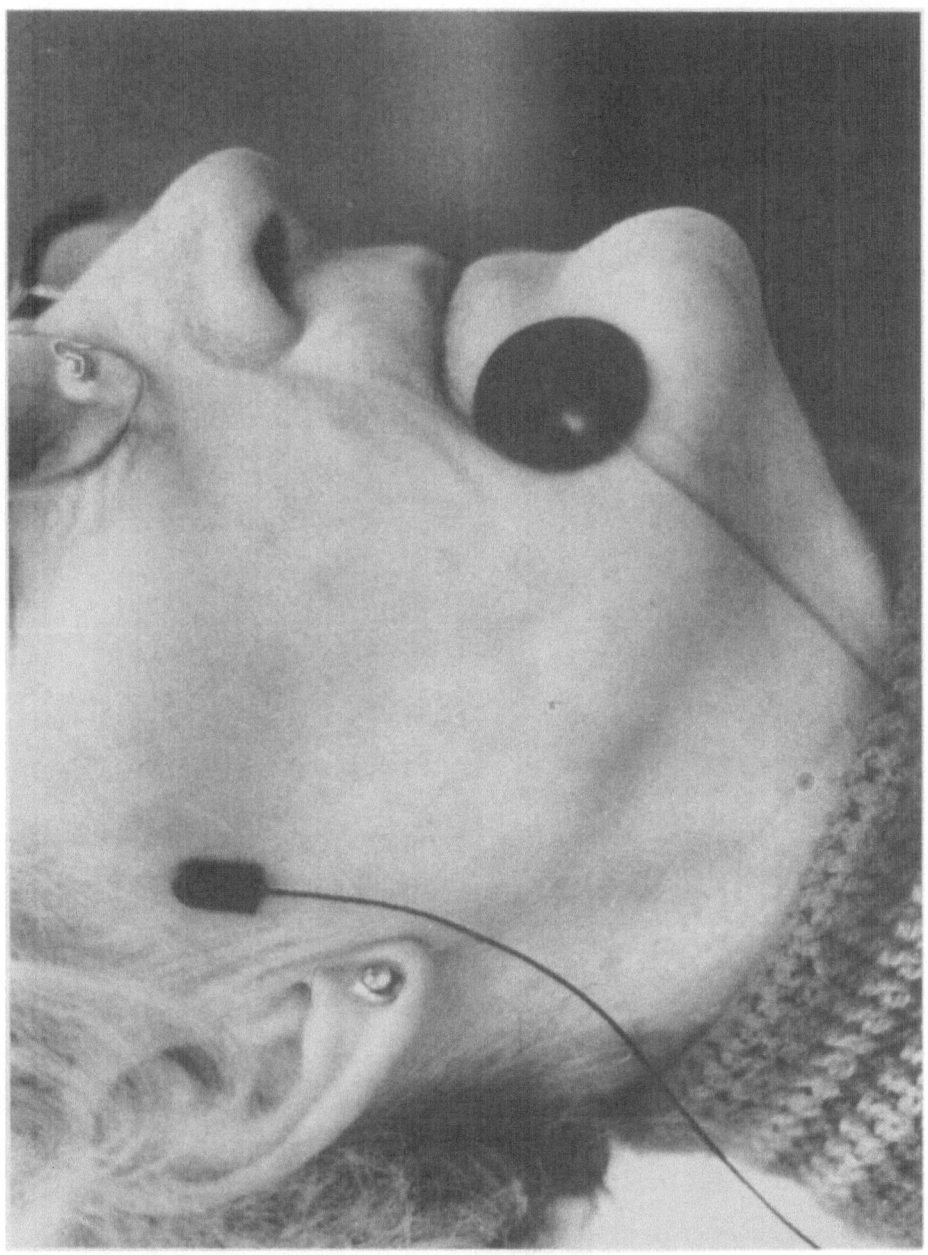

Fig. 40: Electrodes in place for treating trigeminal neuralgia (third branch on right side; small electrode = anode and cathode over exit of mandibular nerve on lower jaw). From JENKNER [118].

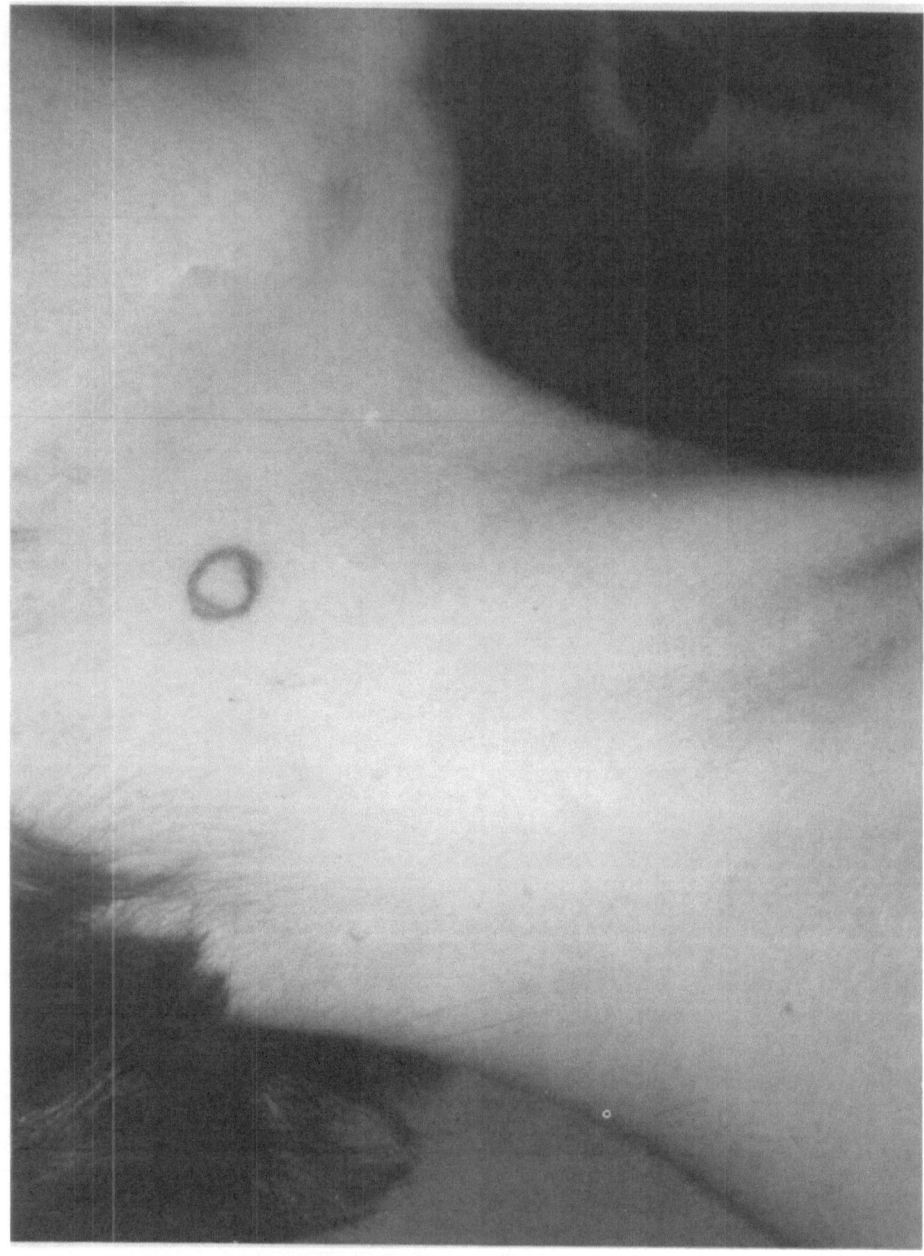

Fig. 41:   Position of anode marked for therapy of right glossopharyngeal neuralgia using special brief DC-pulses. Cathode on opposite side and of the size of a cathode for trigeminal neuralgia.

CERVICAL PLEXUS and BRACHIAL PLEXUS

Segmental Access: The cervical plexus has afferent and efferent fibers from the segments C-1 to C-4, brachial plexus those from C-5 to C-8 and D-1. For correspondence of pain areas and segmental innervation see "INDICATIONS".

Anatomy: All cervical nerves leave the spinal cord, pass through the intervertebral foramina and continue their course laterally passing through the indentation between the anterior and posterior tubercula of the transverse processes of each cervical vertebrae. At this site they run very closely under the skin and directly over a bony landmark, which makes them easy to identify. Pressure on the nerve at this point against the bony mark will inevitably produce pain sensation irradiating into the respective area of fiber distribution of the nerve. Before their definite course to the periphery of the body, they are situated side by side, exchanging fibers at specific stretches thereafter, forming the cervical and brachial plexus. Continuing towards the periphery, they supply the head from C-2, neck and collar from C-3 and C-4, and shoulder region from C-5; certain fibers go down the arm (from C-6 to D-1). Exact correlation of pain areas with supply areas of cervical nerves are to be found under "Indications". For proper localization of a certain nerve, it is suggested to start counting from the anterior tubercle of the sixth cervical vertebra, which is the most prominent and is also easy to localize within the borders of the supraclavicular fossa. The other points of localization for the more craniad nerves are to be found along a line from this landmark to the mastoid process. The more caudad nerves C-7, C-8 and D-1 are located much too deep at this anterolateral site and should be looked up from posterior, where they may be found as described under electrode positions.

Positions and Sizes of Electrodes: The nerves C-4 to C-6 are situated very closely under the skin at the above mentioned site, where the effect of electrical impulses is excellent. This is important to note since these nerves are most often involved. The position of the anode is exactly over the spot described under "Anatomy" for the nerves C-3 to C-6: between the two tubercles of the transverse processes of the respective cervical vertebra. The size of the anode should preferably be about 7.3 cm^2. For these nerves, the cathode should be 96.8 cm^2 in size and is to be placed over the dorsal processes of the cervical vertebrae and the two uppermost thoracic vertebrae. The direction of the longitudinal axis of the cathode should be the mid-sagittal line. This position is correct for either side and also as a common cathode for bilateral application of anodes, even if not in the same but in adjacent segments. For the segments C-7, C-8 and D-1, the small anodes (size as mentioned above) should be placed in a transversal line through the uppermost margin of the C-7 dorsal process for the C-7 segment, the lowermost border of the C-7 process for the C-8 nerve, and the uppermost border of the first thoracic dorsal process for the segment D-1. The placement should be about 3 cm (1.2 in) to the involved side, but this distance should be corrected for variations in skeletal built of the patient. The cathode for these nerves should be placed over the mid third of the clavicle with a long axis parasagittally. Here we cannot use a common cathode for bilateral treatment, but a separate cathode has to be placed on each side. The size of the large cathode should be as mentioned previously.

INDICATIONS:

REGIONS OF PAIN	PROCEDURE
Occipital pain up to the border of hair on the forehead, mostly without pain on pressure of greater occipital nerve.	Block of C-2, preferably lower anodal position over greater occipital nerve (see p. 81).
Pain in neck and collar from border of face at lower jaw to Adams apple.	Block of C-3. Anode laterally.
Pain in upper half of clavicle, supraclavicular groove, to Adam's apple and the border of C-3; in the neck, pain craniad of dorsal spine of C-6 to border of C-3. Sensation of a lump in the throat.	Block of C-4. Anode laterally (fig. 43)
Pain in upper part of scapula including suprascapular, in part also infrascapular muscles; acromion, acromio-clavicular joint and lateral part of bone of upper arm; pain superficially on skin of lower arm (volar face) ending in a sharp angle at wrist (misdiagnosed most commonly as "tendovaginitis" and/or "tenosynovitis").	Block of C-5. Anode laterally.
Pain in upper part of lower half of scapula, over lateral part of acromion, medial part of bone of upper arm, head of radius, lateral part of radius, of carpal bones, and of first beam; skin of thumb and radial part of index finger at its base. Vertigo, tinnitus, visual disturbances, migraine, cardiac sensations, variations in blood pressure, hyperhidrosis etc.	Block of C-6. Anode laterally (fig. 44)
Pain in lowest part of scapula, descending on upper arm spirally from medial to lateral, both epicondyli, proximal part of ulna, distal part of radius, bones of 1st to 3rd beam of hand, skin of ulnar face of index finger, whole middle finger and radial part of ring finger. In case of ineffective local therapy of epicondylitis: look at C-7!	Block of C-7. Anode dorsally (fig. 45)
Pain of most distal volar part of upper arm, often together with D-1 segment, distal part of ulna with bones of 4th and 5th beam of hand; skin of ulnar face of ring finger and whole little finger and styloid process uf ulna.	Block of C-8. Anode dorsally
Pain of lower arm and hand (all nerves) from any cause whatsoever. For differentiating central from peripheral and functional from organic pain.	Brachial plexus block at medial face of upper arm (placements of anode see photo). Cathode on lateral face of upper arm.

Intentionally, nothing has been said so far about the actual causes of pain. Pain from all causes may be treated, as long as it can be localized to the segmental symptoms described above, also in the case of malignancies. Pain of overlapping segments or extending over several segments or differing sides may be treated by placing anodes over all of the respective nerves.

This only applies, if pain is *not* due to malposition of the cervical spine. If there is such a malposition, only adjusting this position to normal (=lordosis) will be of help [119].

Readjusting the position of the cervical spine requires (1) a clear cut diagnosis on level and kind of malposition (96% a kyphotic kink, 2-3% a lordotic kink, about 10% in addition to a kink a slight subluxation of the higher vertebral body either ventrally (somewhat more frequent) or dorsally over the lower vertebral body of the involved segment. (2) a totally relaxed patient; which state has to be tested by very slight extension of the cervical spine via the head (right hand on chin, left hand, volar face up on occiput; for right handed physican) and rotating the head about 30⁰ to each side. (3) holding index finger of left hand exactly on dorsal spinous process of lower vertebra of involved segment, steadying it by some pressure and hyperextending head and cervical spine to about 90⁰ if patient tolerates. If smaller extent of hypertension is tolerated only, this smaler degree should be enough. And finally (4) keeping chin in place and lifting occiput up to normal mid-position, the position which was the starting position; then (5) slowly releasing the minimal degree of extension (which might be a bit painful in an acute case or at first manipulation). – This procedure should be repeated once a week on same day and after 3 maneuvers of readjustment one week should pass: control by lateral x-ray film should show a return to normal lordosis. Explanation for lordotic kink or subluxation shall not be given here. The reader is referred to JENKNER [119, 121].

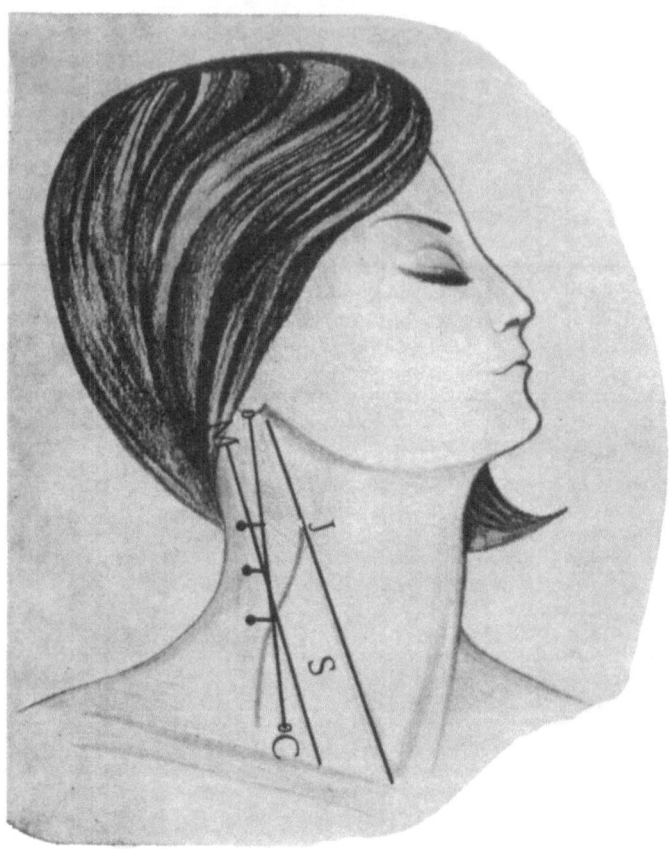

Fig. 42: Anatomic sketch for finding position of anodes to electrically block the cervical nerves. From JENKNER [118].

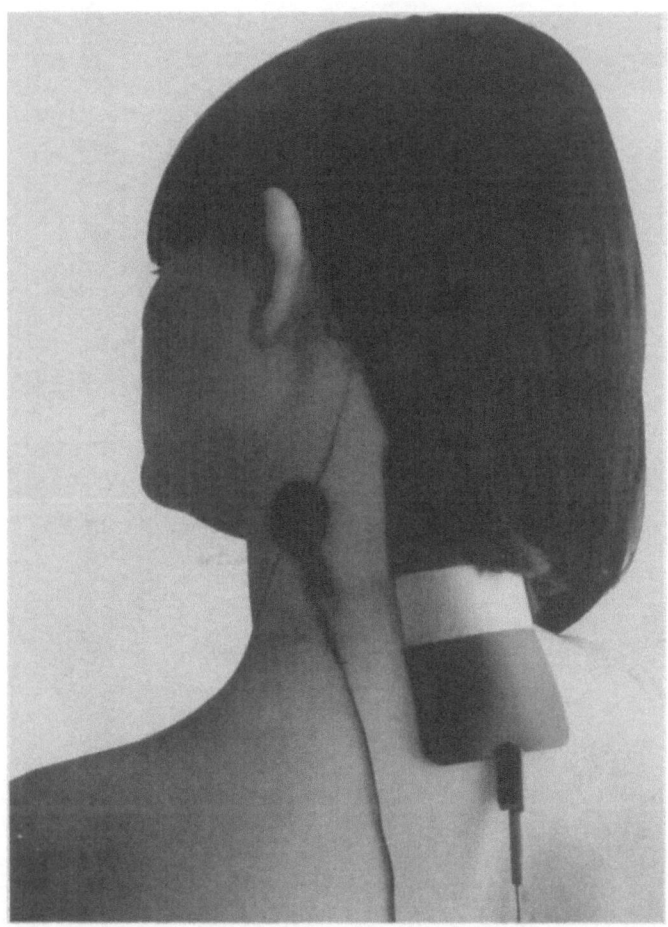

Fig. 43: Position of small anode and large cathode for treating pain in area of 4th cervical nerve on left. Full line indicates connection between mastoidmprocess and carotid tubercle, on which all anodes should be placed.

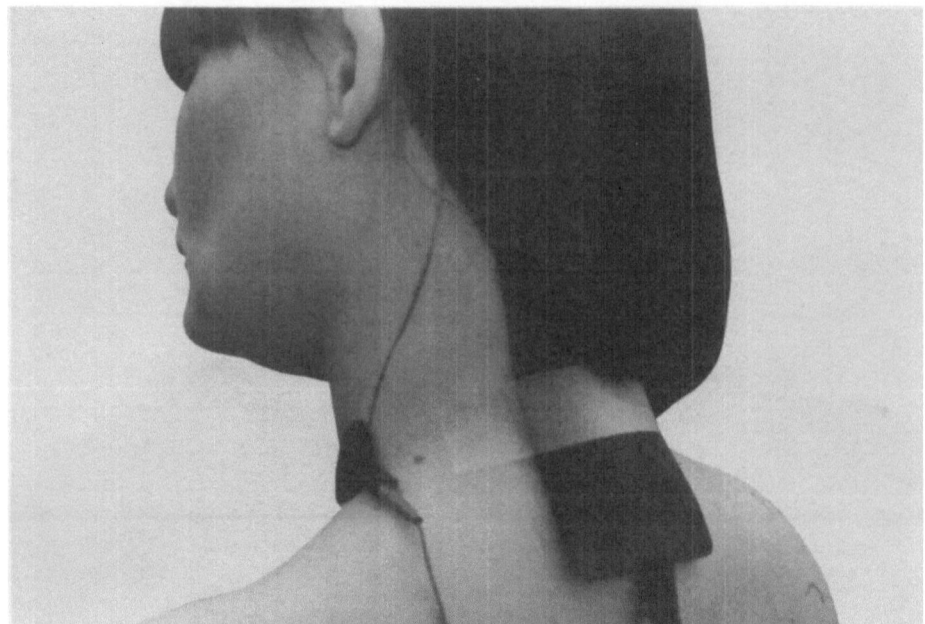

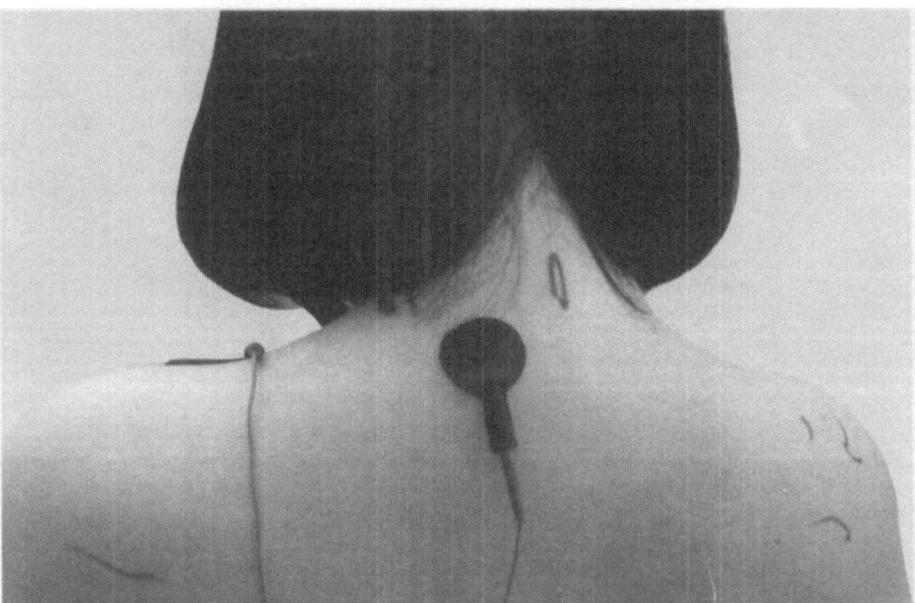

Fig. 44: Position of round anode and rectangular cathode for electric block of C-6 left. Line as in fig. 43.

Fig. 45: Position of small anode and large cathode (of which only upper margin with cable connection is to be seen) for blocking the 8th cervical nerve on left. The marked oval craniad of electrode indicates contour of tip of 7th dorsal spinous process. The same distance from, but craniad of the oval the anode would have to be placed for therapy of 7th cervical nerve using same cathode position.

SUPRASCAPULAR NERVE

Segmental Access: Fibers for this nerve originate in the segment C-5 only.

Anatomy: Just as is the case for blocks with a needle, this nerve should be reached by the electric field at its exit through the suprascapular notch. To find this site easier, please refer to figure 46. The lines shown in this figure are marked on the skin as shown in fig. 47. This helps in daily use of this block to determine the correct anode placement.

Positions and Sizes of Electrodes: A 7.3 cm^2 large anode should be placed over the point where the lines shown in fig. 47 intersect. Fixation ist best assured by karaya. The cathode, 96.8 cm^2 in size, should be placed opposite the anode on the skin of the ventral surface of the body with its longitudinal axis in a parasagittal line. These positions may be seen in figures 47 and 48.

Indications: This block is of value in differentiating the various causes of shoulder pain and pain high up in the back. However, one should not forget that the scapula is innervated by three nerves and the most cranial one is the 5th cervical nerve. Therefore, this block will only be able to influence pain in the upper half of the scapula. The lower half of the scapula is innervated via the 6th and 7th cervical nerves. The block of the suprascapular nerve will help especially for pain due to a tunnel syndrome at the notch, in which case this point is extremely painful to pressure from the skin. It is of use also in pain from subacute and chronic subacromial bursitis (only as an adjuvant procedure, not an adequate therapeutic measure by itself). In cases of humero-scapular periarthritis, its pain reducing effect should be supported by active and passive physiotherapeutic exercises. Sometimes pain at the point of the notch caused by a cervical syndrome also may be reduced by this block.

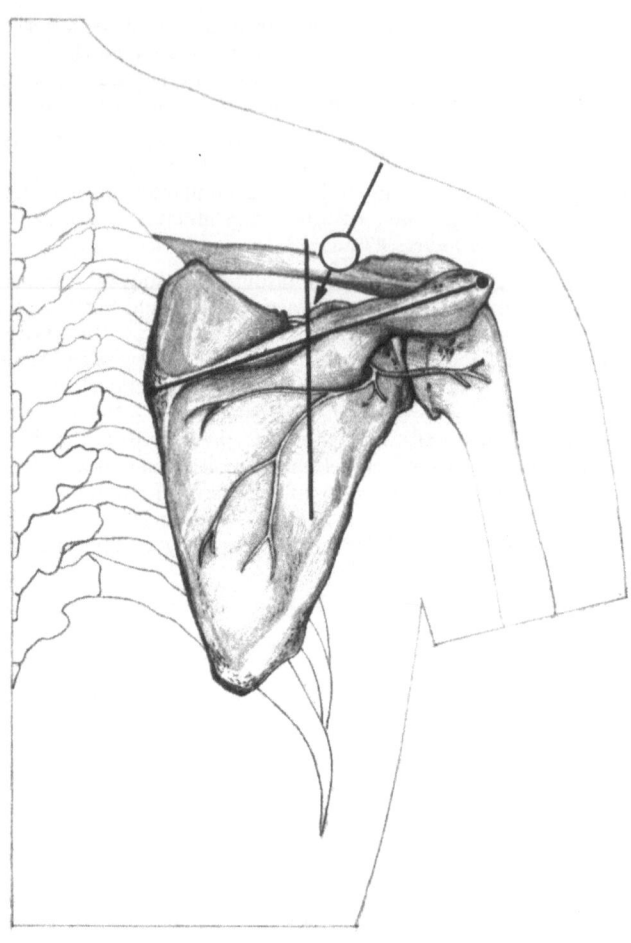

Fig. 46: Anatomic sketch to find position of anode for electroblock of right suprascapular nerve. From JENKNER [118].

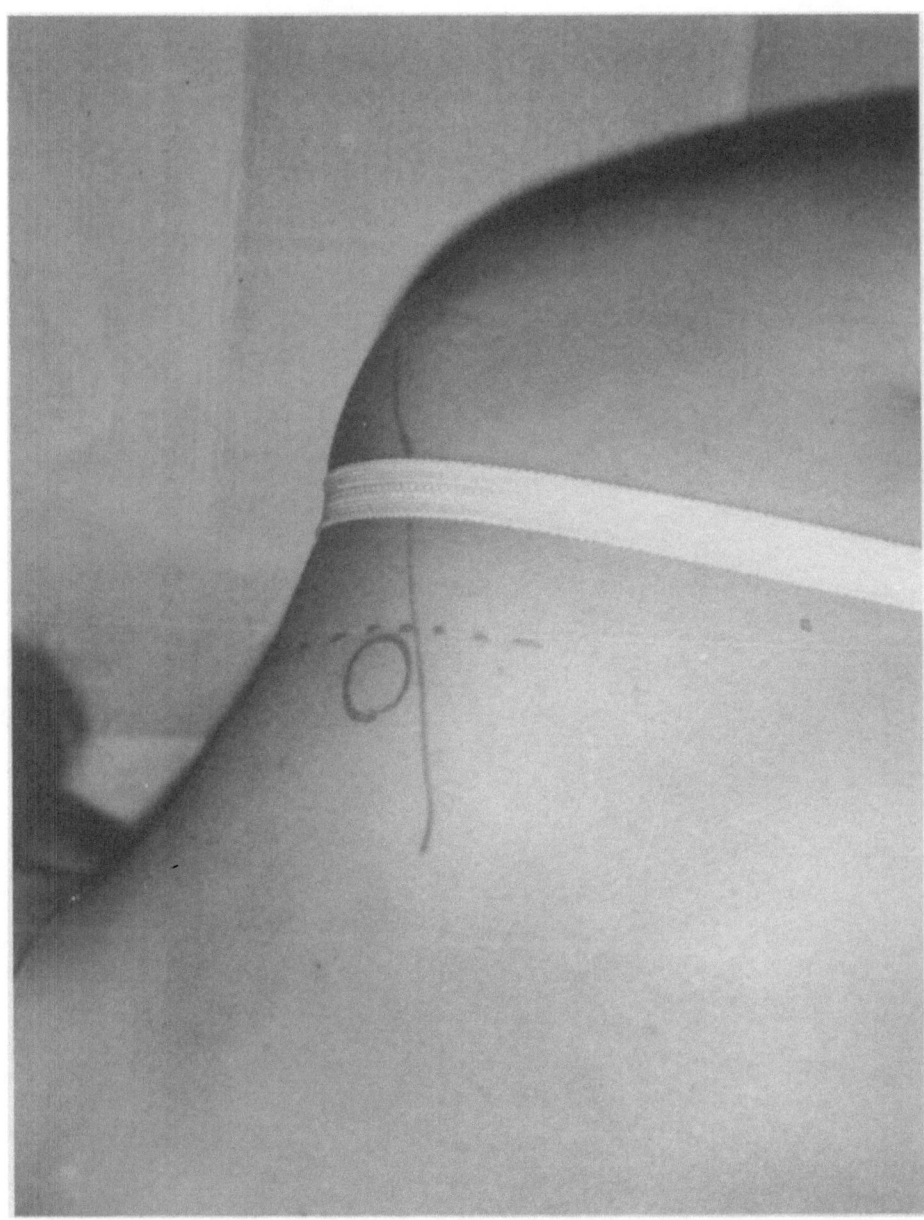

Fig. 47: Method of marking skin to arrive at anode position for blocking right suprascapular nerve. The almost horizontal line indicates the spine of scapula. The
second line is halving the first, but positioning of anode depends on pain the
patient would sense: either in lateral angle or medial angle (as indicated here)
the anode may be placed.

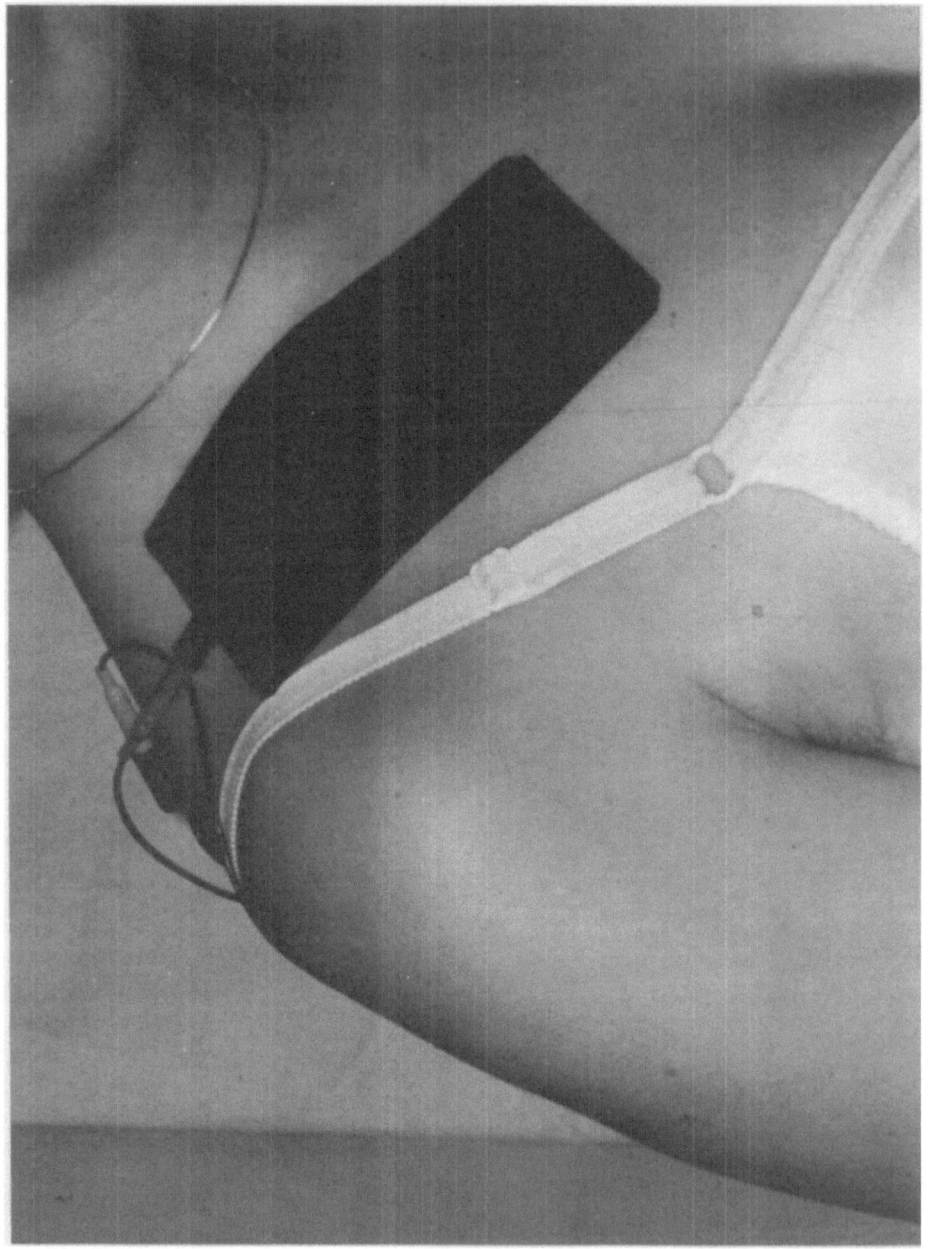

Fig. 48: Placement of cathode for treating pain in area of right suprascapular nerve.

PHRENIC NERVE

Segmental Access: The fibers for this nerve are derived from the segments C-3 to C-5 .

Anatomy: The position of the phrenic nerve in the neck immediately under the sternocleidomastoid muscle (which means relatively close to the skin) is an advantage which makes electric blocking very effective. The exact site of blocking is evident from figure 49. Testing the functioning or malfunctioning of this nerve actually should be done by determining the nerve conduction velocity. However, since this is hardly ever possible in the practical situation, a procedure which is possible everywhere is suggested instead: radiographic imaging on a screen should be carried out to ascertain (and document on film) normal respiratory excursions of the diaphragm (which is referred to as model A below) or diaphragmal paresis (which is referred to as B below).

Positions and Sizes of Electrodes: In both modes of application (A and B), a small (1.7 cm^2) electrode is affixed with karaya at the correct position 1-1.5 cm (depending on the built of the patient) anterior to the dorsal margin of the sternocleidomastoid muscle 4-5 cm craniad of the upper margin of the clavicle (see fig. 50). The large (96.8 cm^2) second electrode has to be placed exactly on the opposite side of the neck with its longitudinal axis slightly oblique corresponding to the direction of the course of the phrenic nerve. For model A, the small electrode is to be connected to the apparatus as the anode, and the large one as the cathode. For model B, the polarity has to be reversed, i.e. the small electrode is used as cathode, and the large electrode should be the anode.

Indications: To prognosticate the effect of a planned surgical sectioning or crushing of the phrenic nerve; in most instances for persistent hiccup. But sometimes, this electric block (model A) is of permanent efficacy by itself, and the surgical procedure need not be carried out at all. In case of diaphragmal unilateral paresis (of not more than 4 months standing), the nerve needs to be stimulated (model B), which will be successful if the function of the nerve was not entirely lost. Electric parameters for both models are identical, i.e. very short (below 200 μ s) single impulses of the nature of true square waves. The number of pulses per second differs; for model A, 20-50 pulses/sec are required and we propose to choose 35/sec (easy to adjust by observing the flicker of the control lamp. It disappears at about 35 pps, which corresponds to the flicker fusion frequency of the human eye). For model B, the optimum would be a pulse frequency of 100/sec or more, but 35 also are effective in stimulating the nerve, even though it takes more sessions to achieve lasting results. One month of daily sessions of 20 minutes each are required in this case. For model A, 5 daily therapeutic sessions are adequate. In both cases, one should verify the effectivity of the therapeutic approach by radiologic controls. Examples of such controls are presentend in figs. 51 and 52.

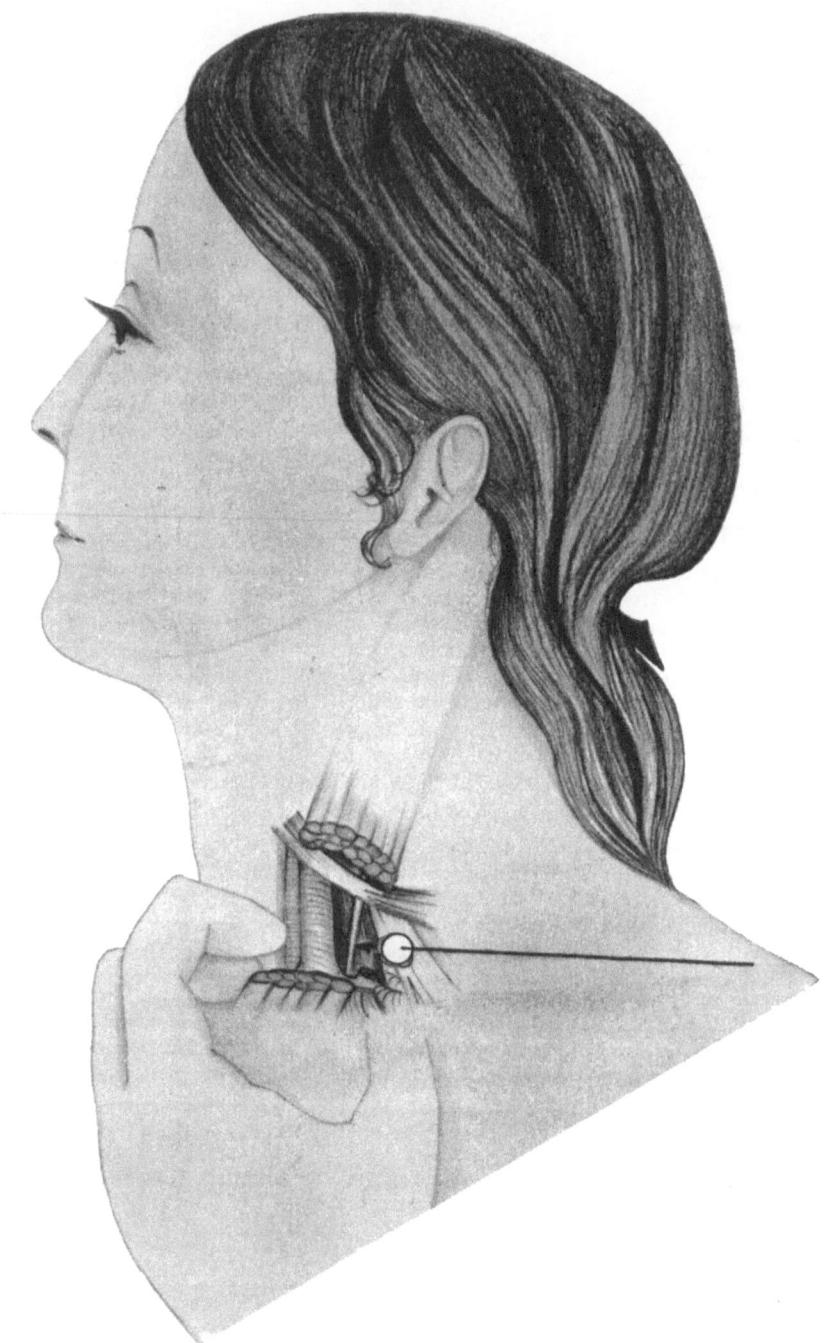

Fig. 49: Anatomic sketch indicating position of left phrenic nerve (arrow) und sternocleidomastoid muscle. Position of anode shown as ring for therapy of singultus.
Slightly changed from JENKNER [118].

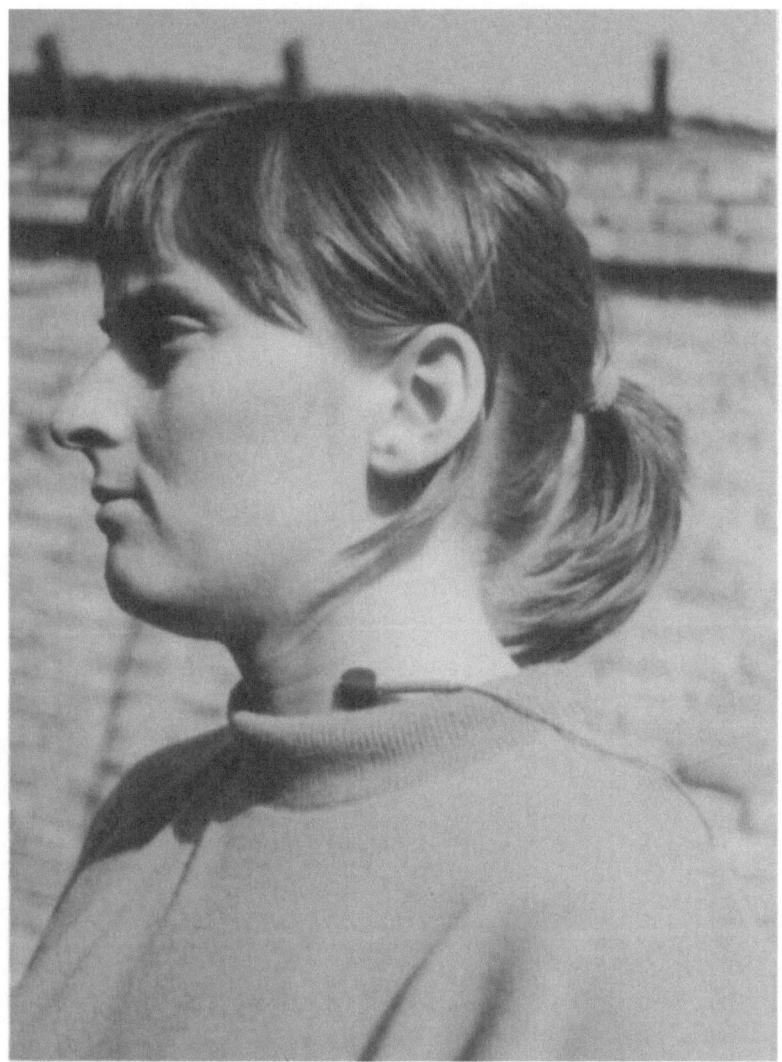

Fig. 50: Photographic documentation of position of cathode for stimulating the phrenic nerve in a case of unilateral phrenic paresis. See also radiologic evidence shown in figs. 51 and 52 (same patient).

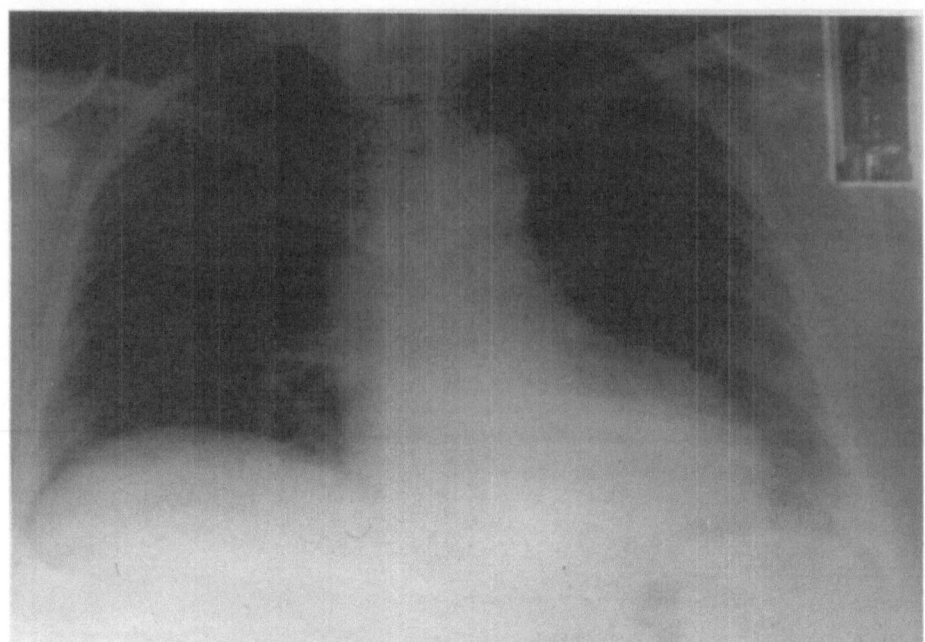

Fig. 51: Radiologic evidence of right sided paresis of phrenic nerve [128].

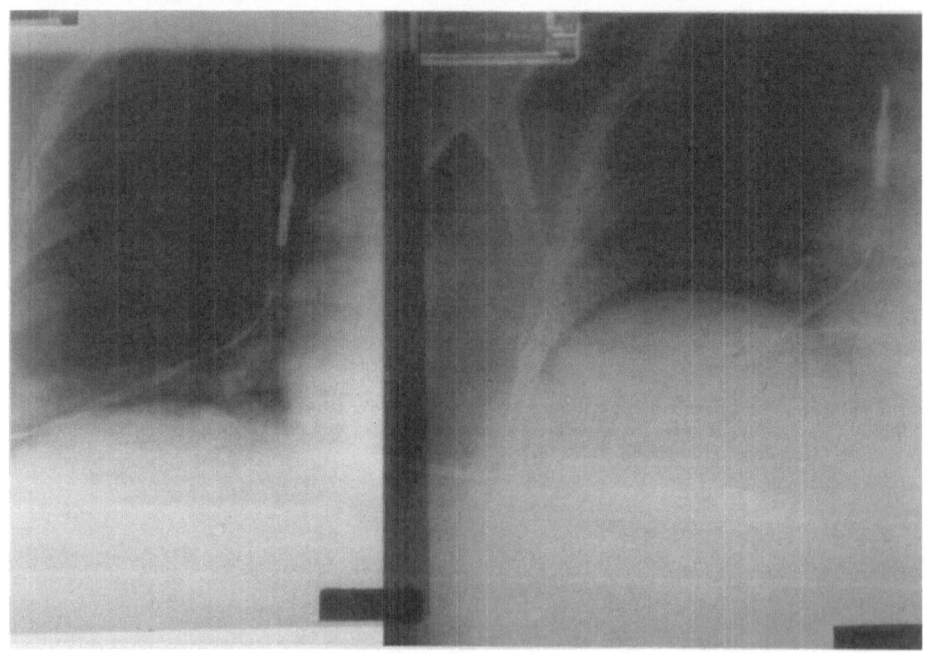

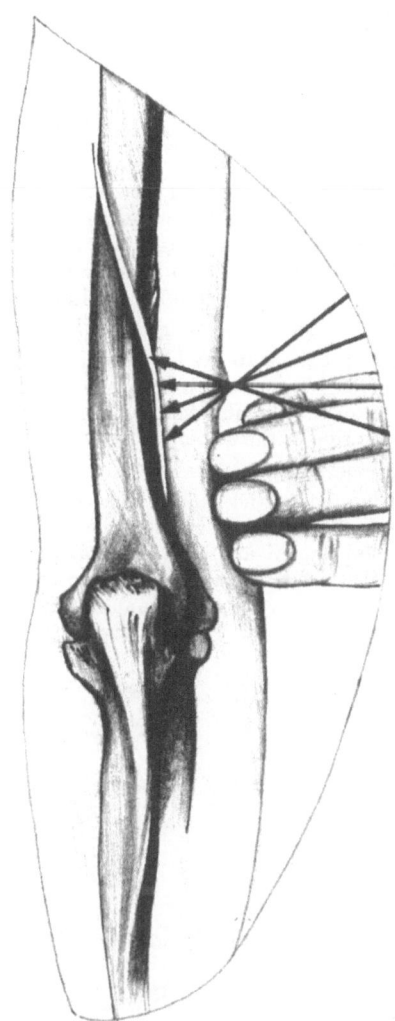

Fig. 53:   Anatomic sketch helping to place anode for electric block of radial nerve on right upper arm correctly. From JENKNER [118].

Fig. 52:   Radiologic observations of testing effect of electric stimulation of phrenic nerve (shade of electrode visible). This therapy allows diphragmal motion (left: exspiration and right inspiratory movement) which without such simulation (see fig. 51) had not been possible. From JENKNER and Fassl [128].

Segmental Access: Fibers from the roots of C-5 to C-8 and D-1 contribute to the formation of the radial nerve.

Anatomy: After its formation, the radial nerve crosses the dorsal face of the humerus to reach the lateral face of it about 2.4 to 3 inches (6 to 7.5 cm) proximal to the lateral epicondyle. There it may be palpated very easily and may be exposed successfully to an electric field (fig. 53). If pains are limited to an area peripheral to the wrist, the nerve may be reached at the root of the thumb (see fig. 54) laterally .

Positions and Sizes of Electrodes: An anode of 0.25 in^2 (1.7 cm^2) has to be placed about 2.4 in (6 cm) proximal to the lateral epicondyle, where it is best attached using karaya as an adhesive. Alternately, in very peripheral pain, it may be placed as shown in fig.55 at the base of the thumb laterally. A cathode of 1.1 in^2 (7.3 cm^2) should be placed over the opposite side of the arm or wrist. The optimum position of electrodes is shown in figs. 55 and 56.

Indications: Pain limited to the supply area of this nerve is rather uncommon. Therefore, the electric field will also rarely be used to influence this nerve. Pain of broader distribution occurs more frequently, and then it is advised to place an anode over the brachial plexus at the medial face of the upper arm near the arm pit (as shown in fig. 61). Vasospastic lesions should be treated via the sympathetic nervous system (stellate ganglion).

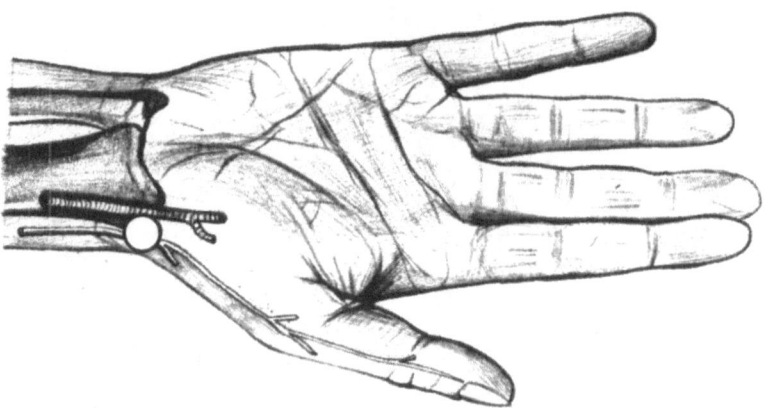

Fig. 54:   Anatomic sketch of position of radial nerve at wrist. From Jenkner [118].

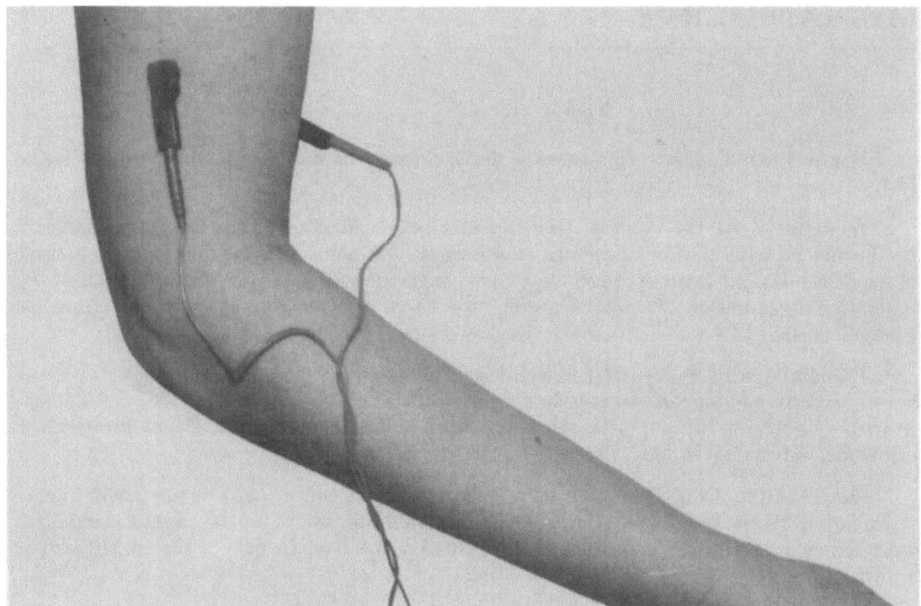

Fig. 55: Position of anode and (partially visible also) cathode at right upper arm to treat right radial nerve there.

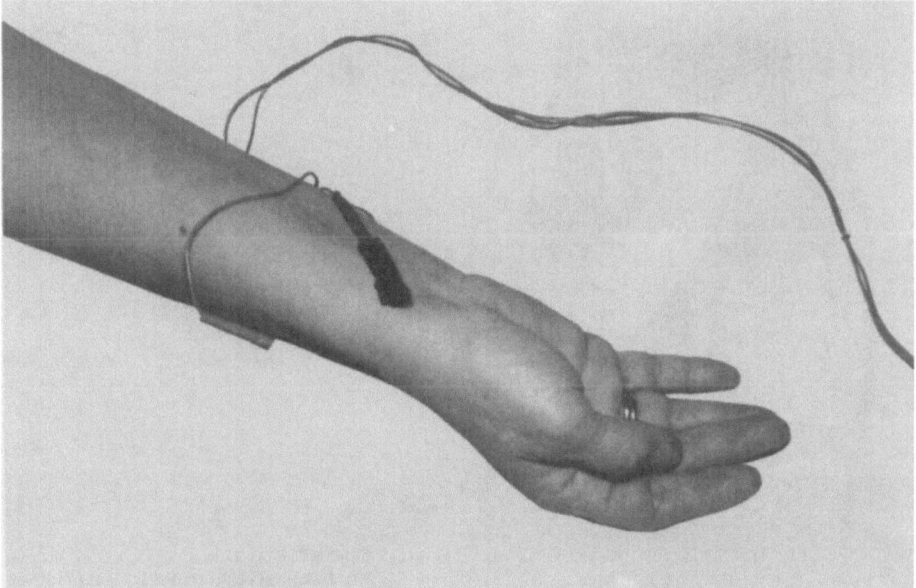

Fig. 56: Position of anode and cathode for electroblock of radial nerve at base of thumb.

MEDIAN NERVE

Segmental Access: This nerve is derived from the cord segments C-5 to C-8 and D-1.

Anatomy: At the cubitus, the median nerve is located just proximal–medially of the fibrous lacertus (easily palpated); at the wrist, this nerve is to be found where a circle through the styloid process of the ulna meets with the radial border of the tendon of the palmaris longus muscle. In both locations, the nerve runs close to the skin, and thus can easily be reached by an electric field via skin electrodes (figs. 57,58).

Positions and Sizes of Electrodes: An anode of 0.25 in² (1.7 cm²) has to be placed exactly over one of the two sites described above. The cathode of 1.1 in² (7.3 cm²) is to be placed on the opposite skin surface of the arm (or wrist). Exact positions of electrodes are shown in figs. 59 and 60.

Indications: Effective only if pain is limited to the supply area of the median nerve peripherally to the position of the anode. Since this will rarely be the case, a more often used anode position is over the brachial plexus on the median face of the upper arm as shown in fig. 61.

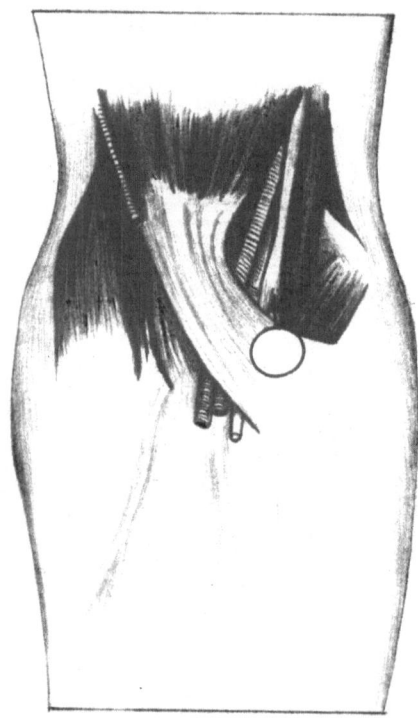

Fig. 57:
Anatomic sketch to find placement of anode for treating median nerve at elbow. From JENKNER [118].

110

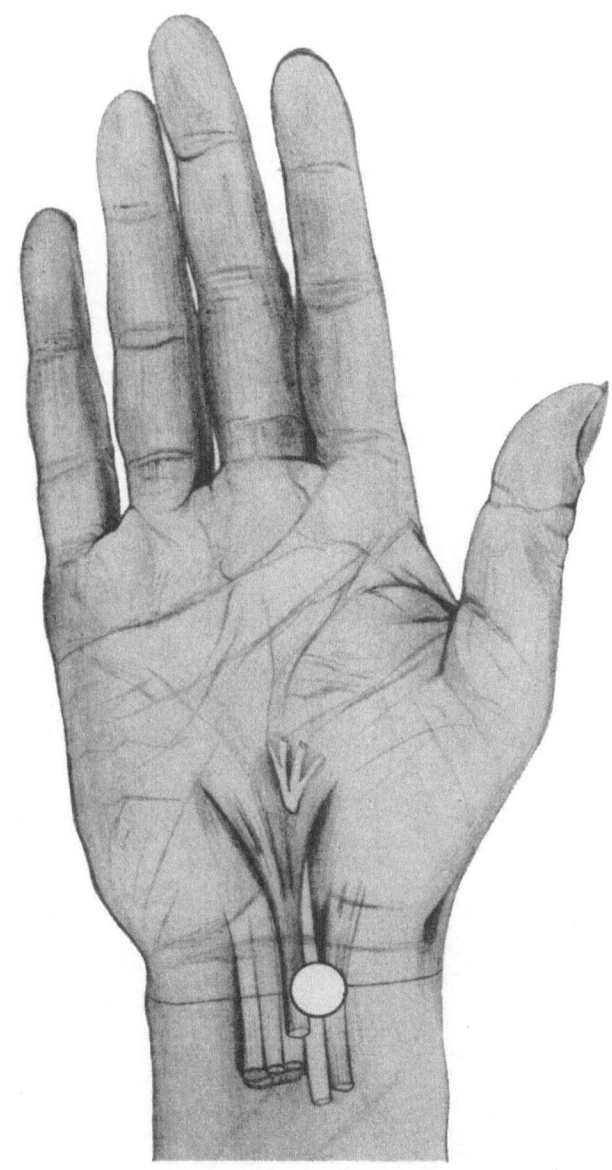

Fig. 58: Anatomic sketch to help place anode over median nerve at wrist. From JENKNER [118].

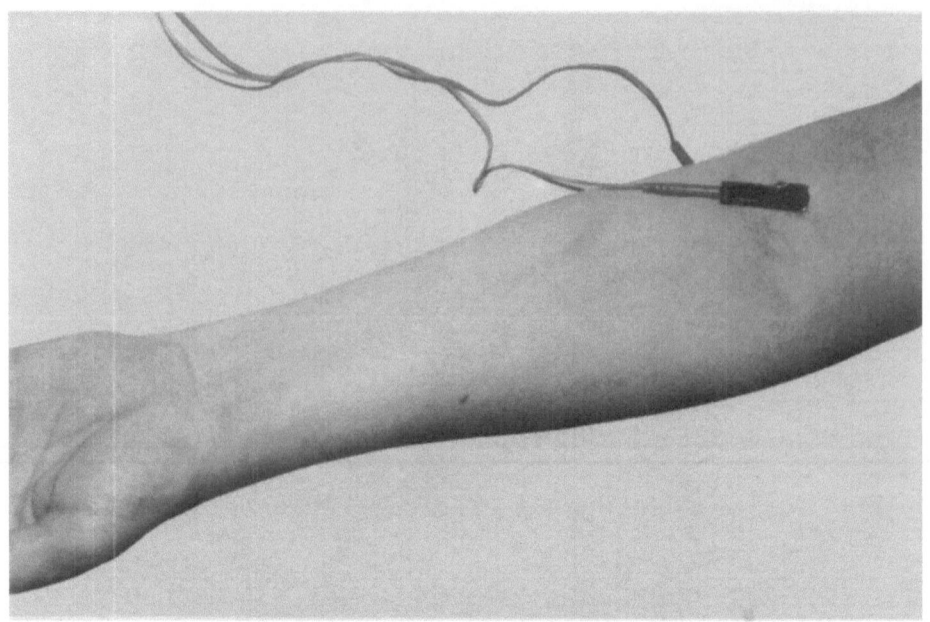

Fig. 59:   Placement of anode for electroblock of left median nerve at ellbow. Cathode to be placed on opposite surface of arm, only cable visible.

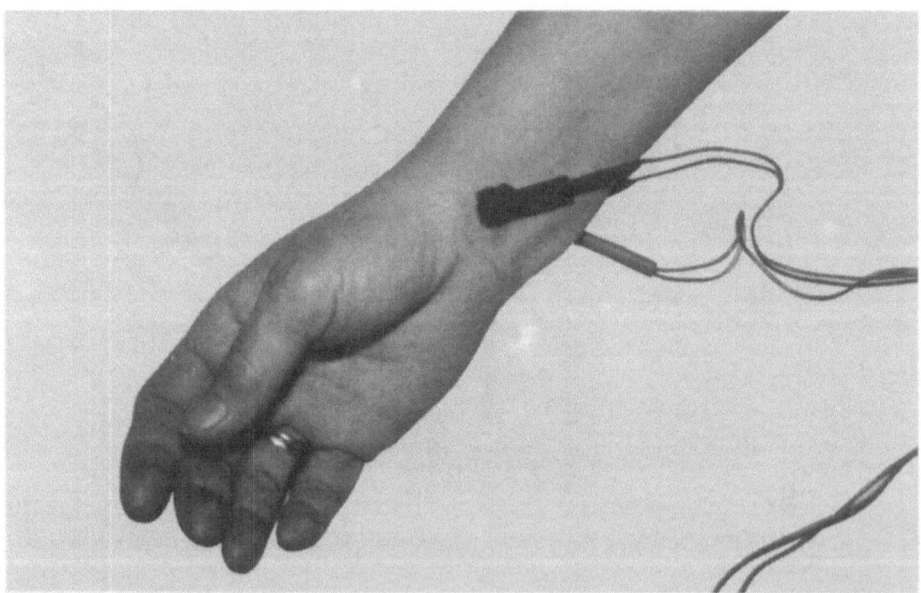

Fig. 60:   Position of anode for electroblock of median nerve at wrist. Cathode to be placed on opposite surface of wrist, only cable visible.

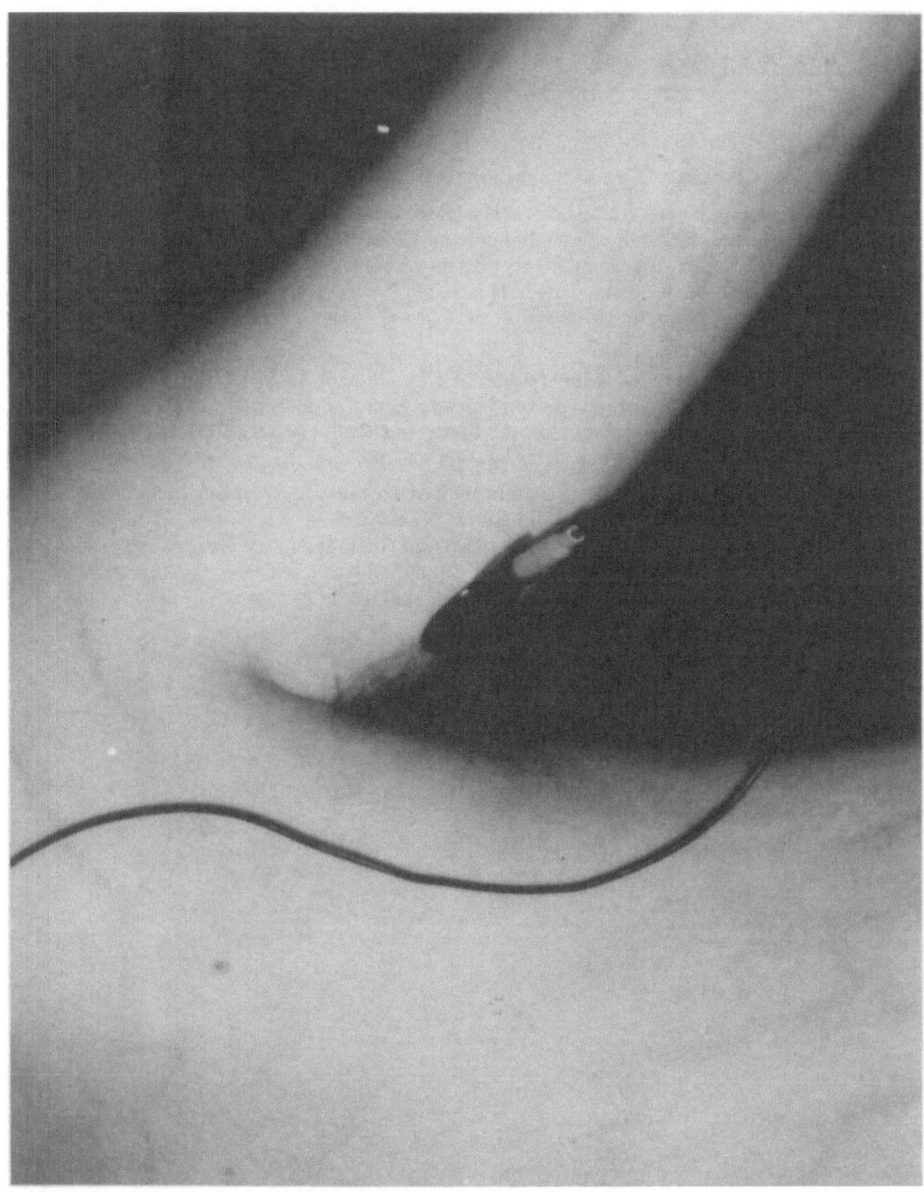

Fig. 61:   Position of anode for electric block of brachial (axillary) plexus at medio-dorsal border of biceps brachii muscle. Place cathode just on opposite skin surface of upper arm with its longitudinal axis in direction of axis of upper arm.

ULNAR NERVE

Segmental Access: This nerve is derived from the cord segments C-8 and D-1 .

Anatomy: On its way from the brachial plexus to the periphery, the nerve passes the ulnar sulcus, situated between the medial epicondyle of the humerus and the olecranon to reach a position between the ulnar artery (on its radial side) and the tendon of the flexor carpi ulnaris muscle on its ulnar (medial) side at the level of the styloid process of the ulna. From there, it branches to the 5th metacarpal bone and the little finger (figs. 62 and 63).

Positions and Sizes of Electrodes: An anode of 0.25 in^2 (1.7 cm^2) is placed on either of the two positions named above, depending on the extent of the pain. A cathode of 1.1 in^2 (7.3 cm^2) has to be placed on the opposite side of the arm (or wrist). The exact position of electrodes may be seen from figs. 64 and 65.

Indications: Effective only if pain is limited to the supply area of the ulnar nerve peripheral to the chosen position of the anode. This case is rather rare, more often the pain diffuses beyond the limits of the supply area of the ulnar nerve (just as is the case for the median or radial nerve). Then, the anode should be applied over the brachial plexus on the median aspect of the upper arm near the arm pit as shown in fig. 61.

Fig. 62:   Anatomic sketch showing site of ulnar nerve at elbow. From JENKNER [118].

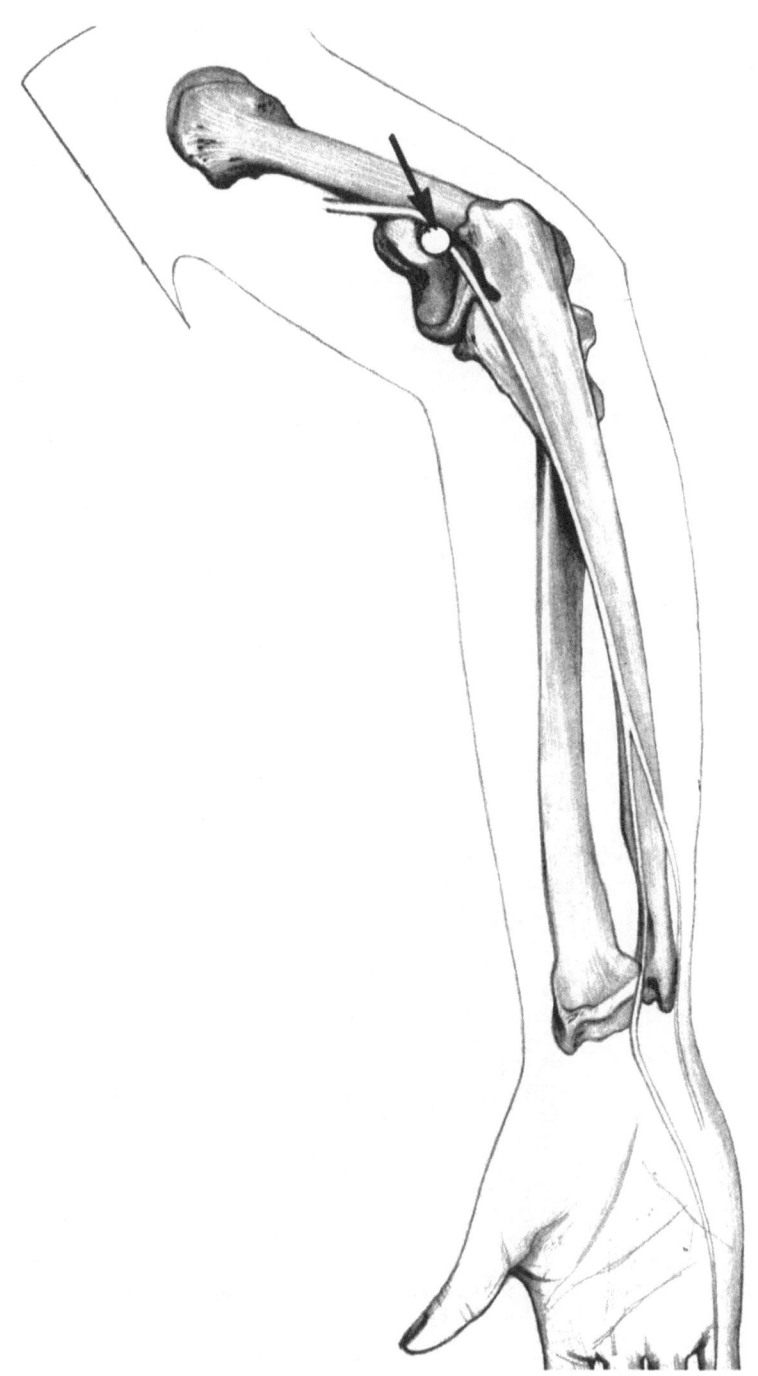

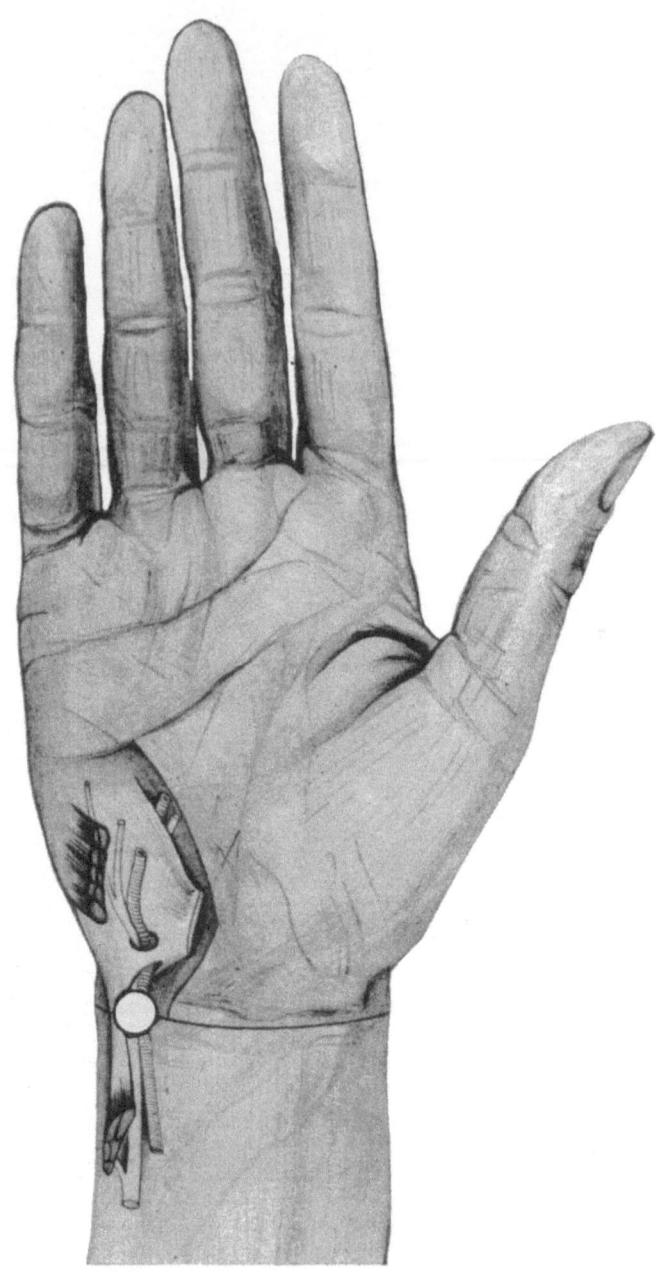

Fig. 63: Anatomic sketch for location of electrode placements to treat ulnar nerve at wrist. From JENKNER [118].

116

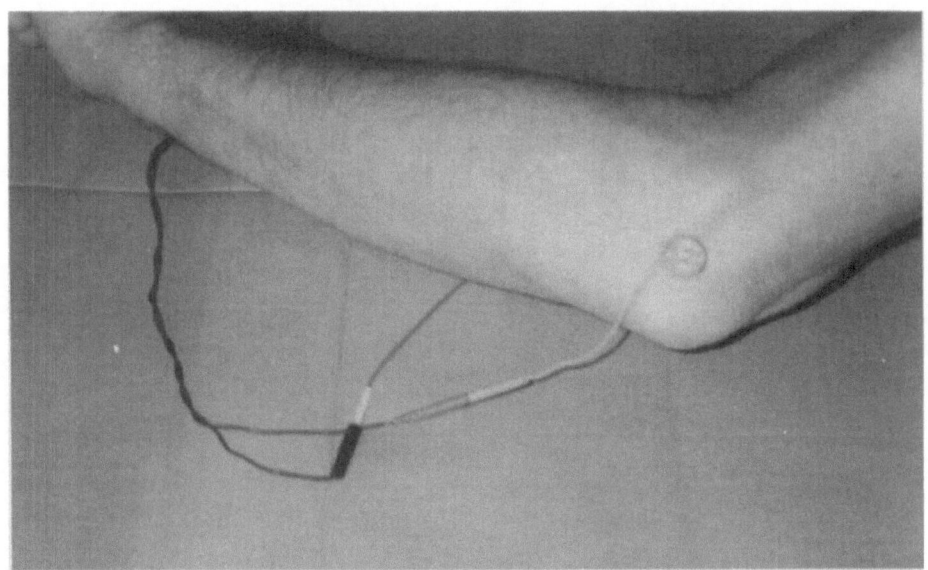

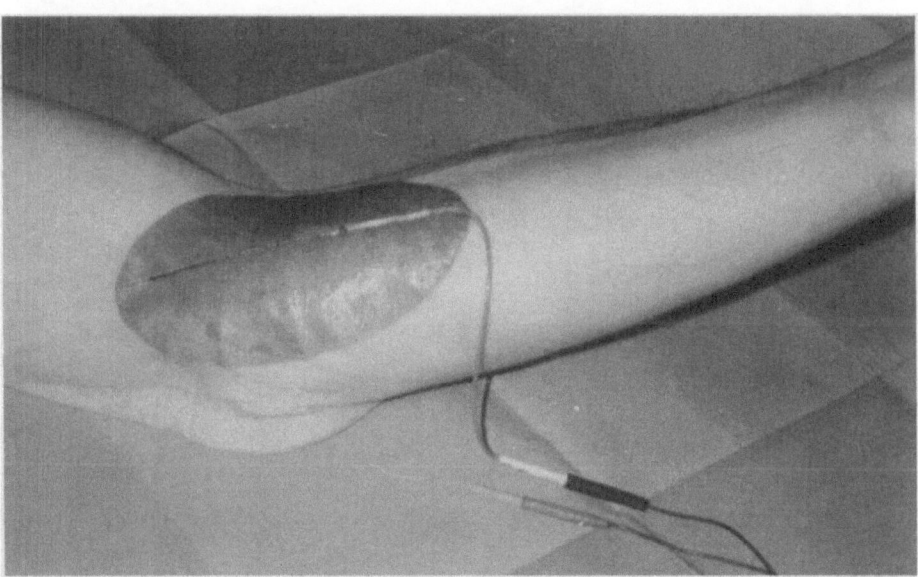

Fig. 64: Position and size of anode placed over ulnar nerve at its groove to block this nerve at elbow (upper photograph). Lower photograph shows position and size of cathode for the same purpose.

117

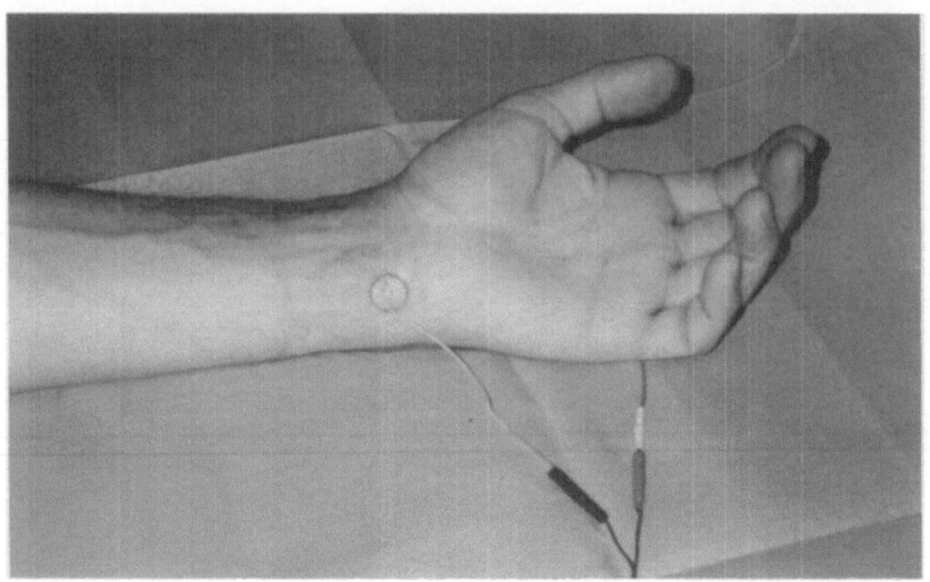

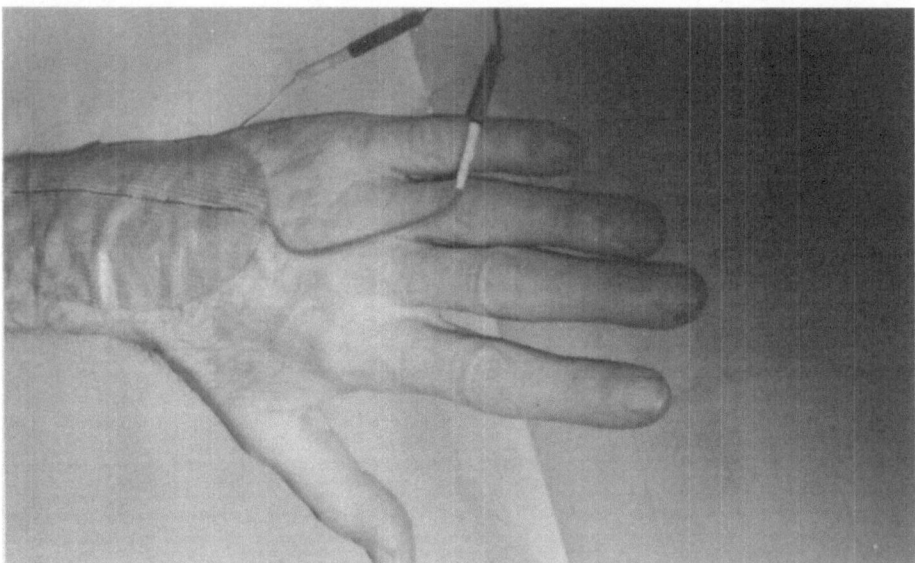

Fig. 65: Position of anode (upper photograph) and cathode (lower photograph) for blocking ulnar nerve at wrist electrically. Polarity has to be changed if stimulation of nerve after injury of nerve is desired.

118

INTERCOSTAL NERVES

Segmental Access: Each segmental nerve obtains its fibers from the respective segment, i.e. from the 1st to the 12th thoracic segment.

Anatomy: Due to the anatomic course and the site of division of these nerves into their branches, there are three locations most suitable for blocking: (1) immediately paravertebrally. (2) at the costal angle not far laterally from location (1). At these two sites, the whole nerve will be blocked. (3) somewhat (about 2 cm or .8 inches) dorsal of the posterior axillary line, where only the anterior branch may be blocked. A block in the anterior axillary line seems rarely advisable; a bit more frequently, a parasternal location is suitable for a block if it is used only to treat pain from sternal fractures and painful pathological processes of the costo-sternal joint. In all cases it should be remembered that the nerve is situated below (caudal) of the rib immediately bordering it. Closer anatomic details, such as are essential for blocking with needles, are not required for electric blocks. But the actual position of the rib and its actual number should be ascertained. This may be difficult in obese and muscular persons, mainly because of the latissimus dorsi muscle and for the first three ribs also because of the serratus anticus muscle; it may even be impossible at times and then one should change to the paravertebral site for finding the correct rib and apply the block there. At this site again, it becomes necessary to recall that dorsal spinous processes of the uppermost ribs are directed caudad, very much overlapping like roof tiles or fish scales (as is known from the cervical spine) and those of the lower-most thoracic vertebrae are directed almost horizontally as is typical for the lumbar spine. The three sites most commonly in use for blocking are shown in fig. 66.

Positions and Sizes of Electrodes: For electric block of intercostal costal nerves, an anode of 7.3 cm^2 should be used and be placed at one of the sites mentioned above. The cathode, 96.8 cm^2 in size should be placed on the skin exactly opposite to the position of the anode, with its longitudinal axis aligned parallel to the dorsal part of the rib. The number of required treatments may vary: in most cases, 5 sessions will bring satisfactory results, except in patients with malignancies, where several months may be required to obtain good results. In such cases, it may be necessary to use 2 daily sessions during the first 3 to 7 days. Such electrode placements for post zoster pain are shown in fig. 67 (anode) and fig. 68 (cathode) for D-3.

Indications: Differentiating somatic from vegetative and mediastinal or visceral pain is possible. Good results are obtained in cases of post-zoster-neuralgias, if antiviral i.v. infusions have been used during the acute phase, the patient did not suffer for more than 2 years and is not over 70 years of age. This Indication is most important, since according to WULF et al. [249] 65% of all zoster infections are located in the thoracic segments. After rib fractures, thoracotomies (of various pathology), upper or even lower laparotomies (postoperative or posttraumatic; in upper laparotomies mainly to ease pain from coughing and to allow better ventilation). Pain from pectoralis minor syndrome, lesions (most often contusions) at the costo-chondral junction and neuralgias late after operative interventions with scars should be responsive to this electric blocking. Intercostal neuralgias of various causes will be relieved if only the exact segmental diagnosis is known. An example for this is shrinking pulmonary lesion with resulting distortion of the thorax and consecutive intercostal neuralgia. Pain from malignancies such as bronchial neoplasms, pleuraendotheliomata, melanomata etc., primary as well as secondary tumors (such as vertebral, rib or pulmonary metastases) as long as they are restricted to a few segments, is treatable for pain relief by electric blocks with good results. The same applies to skin

119

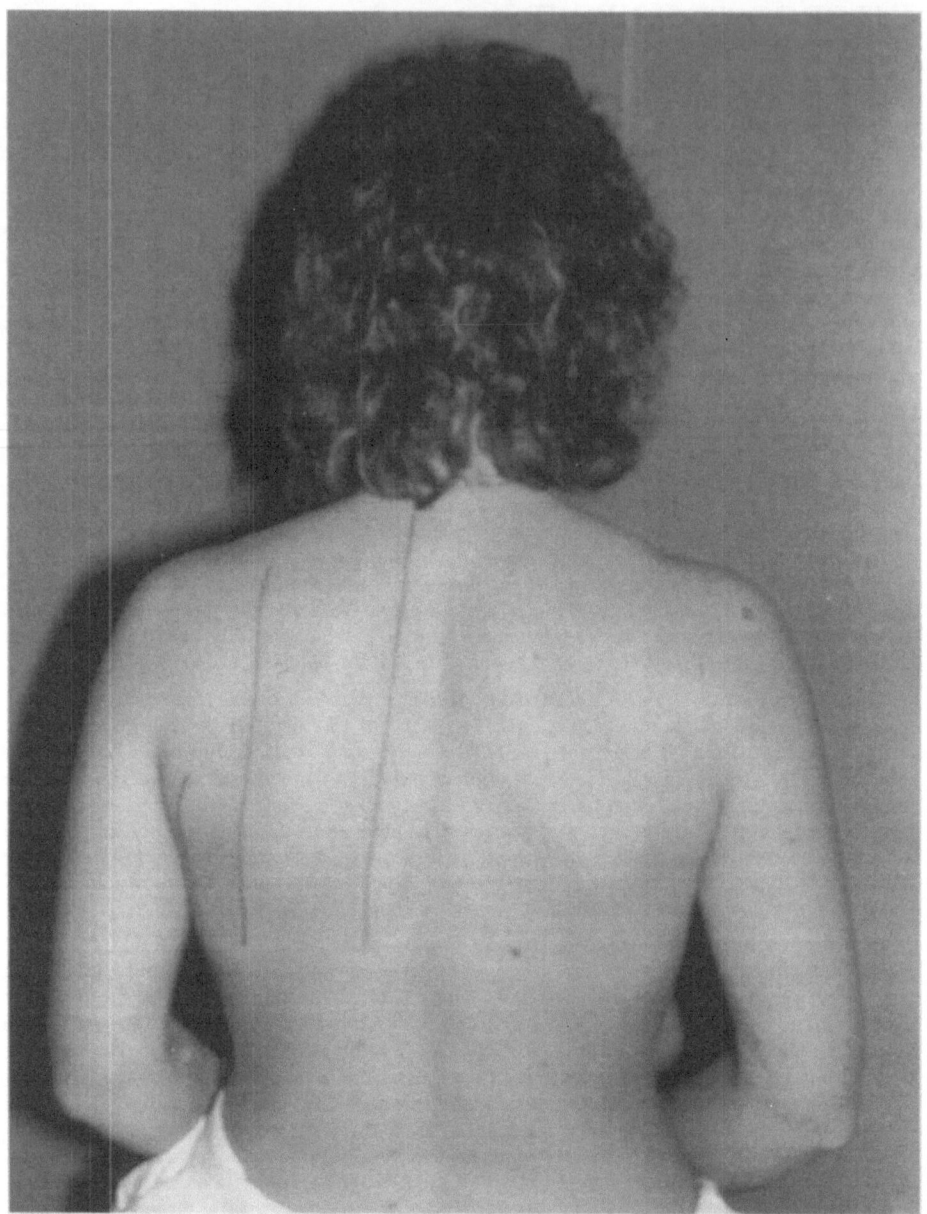

Fig. 66: On chest of a voluntary subject those three lines were marked which allow to locate the 3 most frequently used sites of blocks for intercostal nerves. On left side of back, from right to left, the lines mark a (1) paravertebral site; (2) position over the capitulum (head) of a rib and (3) site 2 cm (=.7 in) posterior of posterior axillary line.

metastases as long as there are not too many. All these lesions may also be situated in or within the abdominal wall since the segmental level of the umbilicus is the 10th thoracic nerve and the small band just above the inguinal fold represents the skin area of the 12th thoracic nerve. Therefore, pain of which the patient hardly will speak of as "thoracic" may be treated by electric block of the intercostal nerves, just as is the case for those painful states of upper back which actually should be treated via cervical nerves, since the innervation area of the cervical nerves is extending down to the lower border of the scapula.

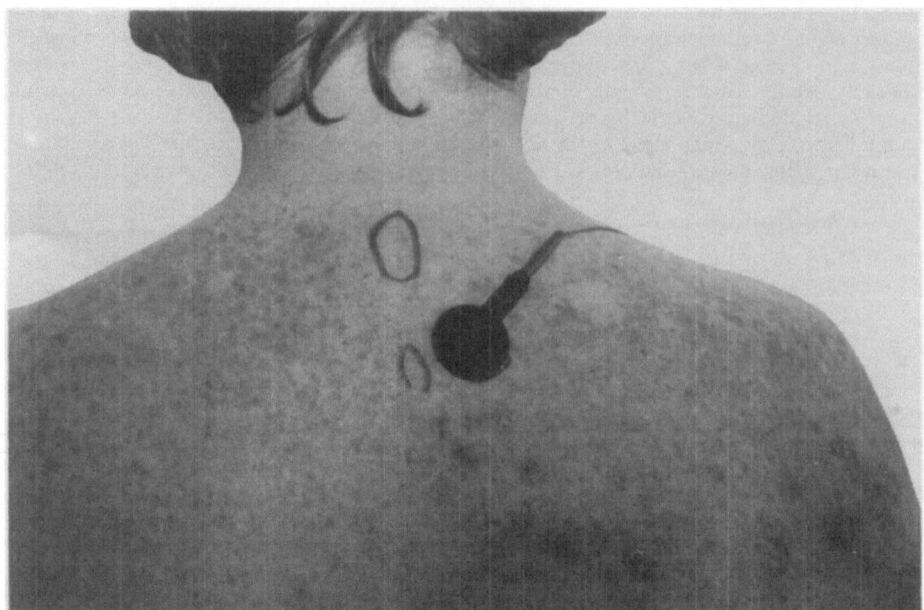

Fig. 67: Position of anode at D-3 right in a case of pain following zoster infection. Oval marks are tips of dorsal spines of 2nd and 3rd thoracic vertebrae. Just over the upper margin of the latter the anode is situated. Dried skin lesions not well visible because of acneiforme skin changes. See also fig. 68.

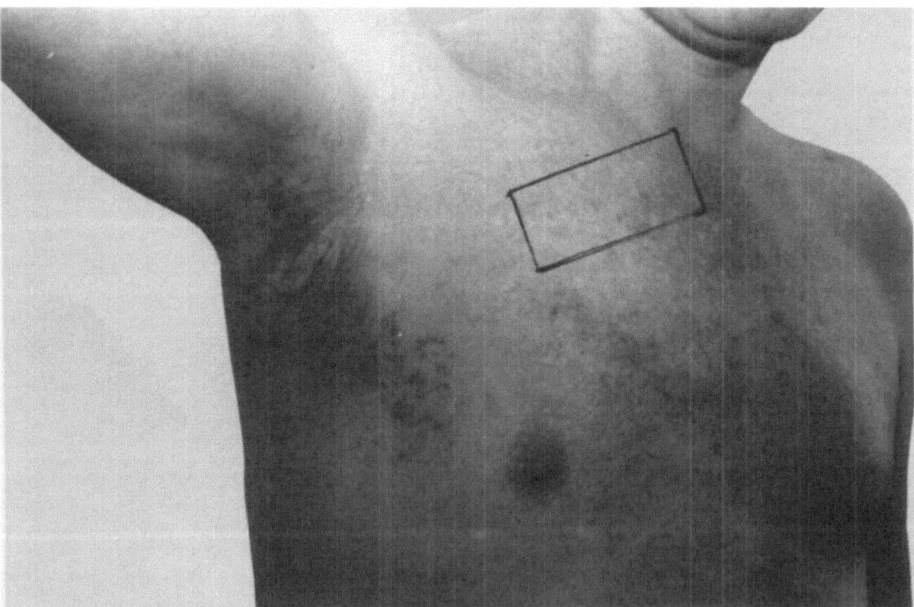

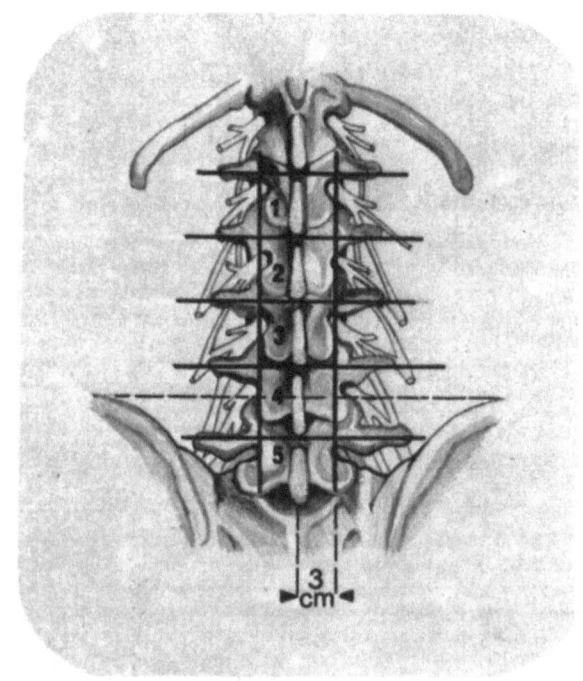

LUMBAR NERVES

Segmental Access: The segments L-1 through L-5 supply fibers to these nerves.

Anatomy:ANATOMY: After passing through the intervertebral foramina they are met best via the horizontally directed transverse processes as landmarks (fig. 69a). The function of these nerves may be checked by certain muscles: m.iliopsoas = L-1 to 4; middle gluteal = L-4 to S-1; quadriceps and adductor muscles = L-2 to L-4; they may also be evaluated by certain reflexes: patellar (tendon reflex) = L-2 to L-4; suprapubic (superficial skin) reflex = D-12; cremasteric (skin)reflex = L-1 and L-2; gluteal (skin) reflex = L-4 and L-5 (see table 5)..

Positions and Sizes of Electrodes: After the respective nerve has been recognized as being responsible for the existing pain, electrodes may be affixed as follows: Anode (7.3 cm^2)should be placed 3 cm (2,5 - 4 cm, depending on bony built of patient and volume of fat) paravertebrally lateral to mid line over exit of respective nerve (fig.69 b). Place cathode (96,8 cm^2) directly opposite of anodal site on ventral face of body homolaterally. Attention should be paid in male subjects to hair since it decreases conductivity and adhesive capacity of karaya - wet pads may be necessary to establish proper skin contact.

Indications: Via these somatic nerves, pain in legs after spinal fusion, vertebral fractures, malignant tumors (either primary or secondary lesions) will be conducted if these nerves are being compressed, pulled or being grown through between cord and periphery by a pathologic process. Visceral pain may be separated from vascular or organic-somatic causes differentiated from visceral-vegetative ones. Pain from kidney, ureteres, bladder and partially from uterus and descending colon have a large sympathetic component and should either be treated via sympathetic pathways or at least be influenced by alternatively using somatic and sympathetic (paravertebral) blocks of the respective segments.

Fig. 69b: Position of anode for electrically blocking the left somatic third lumbar nerve. Cathode is placed on ipsilateral abdominal wall opposite of anodal site.

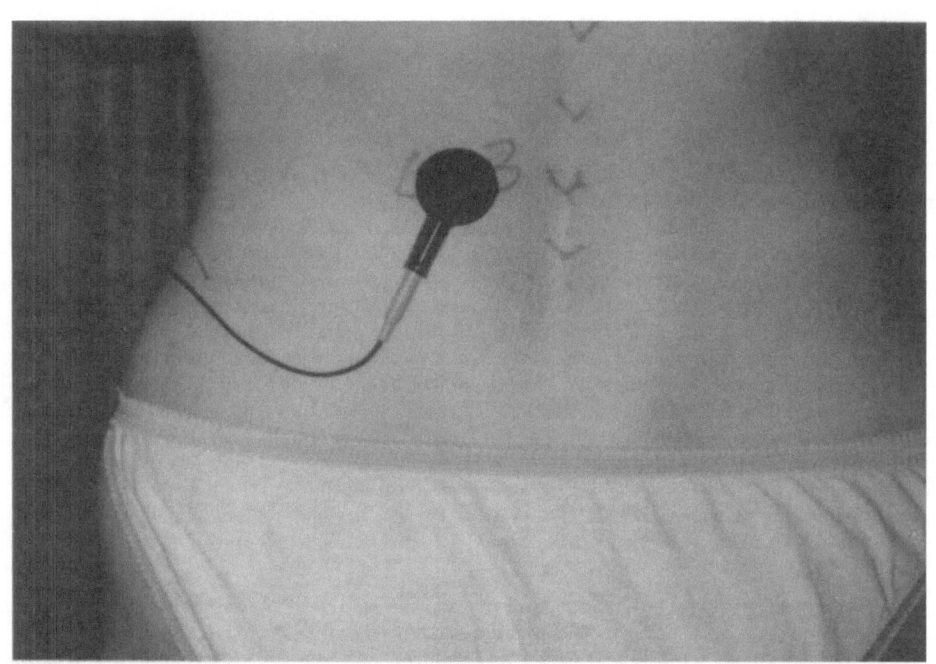

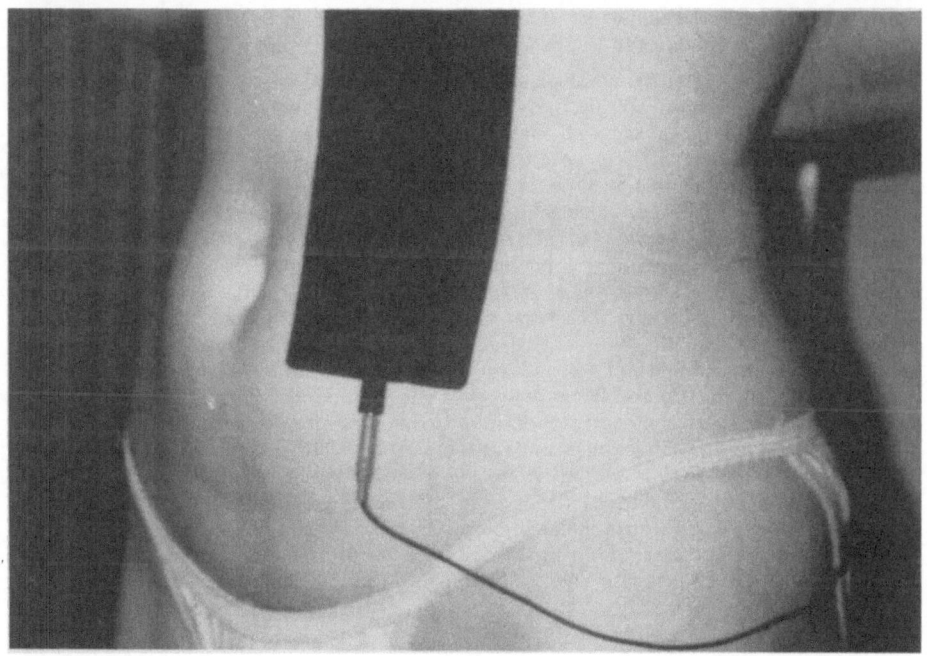

ILIOINGUINAL AND GENITOFEMORAL NERVES

Segmental Access: Fibers from the segments L-1 (the second one also from L-2) make up these nerves.

Anatomy: The ilioinguinal nerve courses subperitoneally around the opening of the small pelvis, runs parallel to the ilio-hypogastric nerve and may not be blocked by itself on a proximal site.It crosses the abdominal wall, follows through the inguinal canal to reach the skin there,where it branches (muscular fibers to abdominal muscles and fibers to the skin of either penis and scrotum or mons pubis and greater labia (fig. 7o). The genito-femoral nerve has his course along the ventral part of the psoas muscle to reach the inguinal canal. There it branches into fibers to the femoral area and genital branches (these only from the segment L-1). These genital fibers reach (after crossing the dorso-caudal part of the inguinal canal) the skin of mons veneris and the labia or scrotum and cremaster muscle. The end branches of both these nerves supply a nearly identical area, one from in front, the other more dorsally. Due to their course, they are predestined to be damaged, contused or sectioned by operative intervention on the bladder, vagina and herniorrhaphies as well as sometimes injuries to the inguinal region. They may – with very minute branches – be tied (compressed) by surgical ligatures and give origin to very painful states.

Positions and Sizes of Electrodes: A 1.7 cm² anode is placed over the peripheral exit of either nerve proximal to the painful site. If this area is free of hair, karaya may be used. Otherwise, a moist pad needs to be placed between the electrode and the skin; but then it becomes most difficult to ascertain good contact since these patients do not tolerate pressure on the skin at all. The 96,8 cm² cathode is to be placed exactly opposite the anodal site on the back of the patient, the patient resting in a supine position.

Indications: After gynaecological operative interventions the greater part of these neuralgiform pains occur. During these procedures small branches of these nerves, almost always invisible to the naked eye, may be irritated or may be sectioned or a ligature tied around it inadvertantly to cause these neuralgias. These painful states seldom occur and are very annoying to the patients concerned since even minor pressure to the skin (garments) is not tolerated. This mainly occurs after vaginal or vesical operations and operations for prolaps or tumors of the uterus. Almost never the cause is readily recognized and most patients suffer a long time before effective measures are approached. These consist in blocking the nerves; the electrical blocks do not molest the patients and are our preferred method of therapy. After one month of daily home treatments pain is greatly reduced so that garments are tolerated again. Only in cases of major strain does pain recur. After another month of daily therapy sessions of 20 minutes each these neuralgias usually disappear entirely and the patients stay free of pain indefinitely. Our control checks the patients once every consecutive month up to one year after the initial treatment series. Then only final evaluation is done. Should the pain be located in the dorsal (not ventral) part of the greater labia, these nerves are not the involved ones but the labial branch of the pudendal nerve (supplied by fibers from the L-3 and L-4 segment via the pudendal plexus, perineal nerve with its labial branch). In these cases the anode has to be placed far dorsally on the greater labium involved. Position of cathode remains identical. The dorsal nerve to the clitoris also belongs to this branch (or plexus).

126

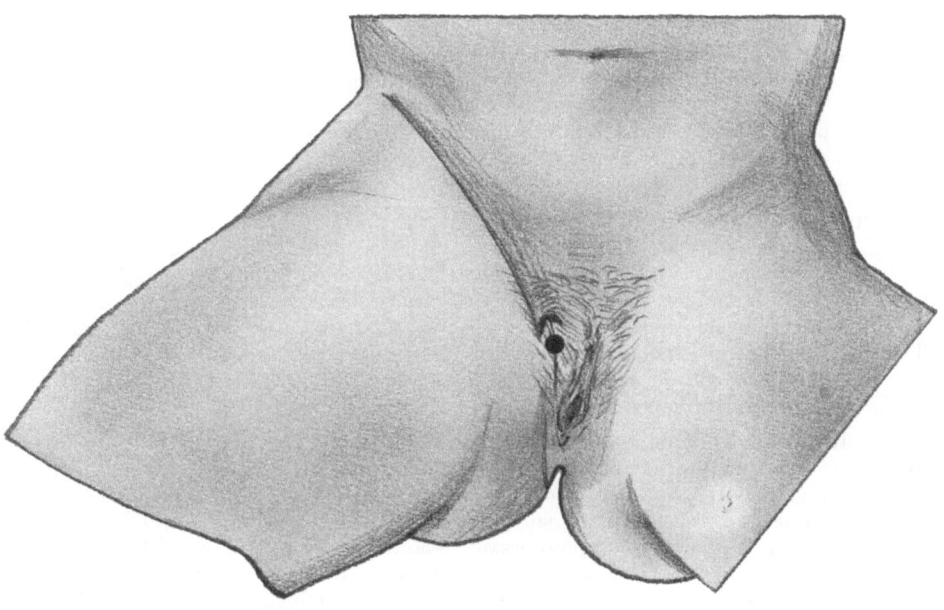

Fig. 70:   Anatomical sketch to find anode position for treating pain in area of ilioinguinal nerve. See text. Anode marked by full dot.

OBTURATOR NERVE

Segmental Access: Fibers of the segments L-2 to L-4 are supplying a branch to this nerve.

Anatomy: Sensory afferents from the hip joint reach the central nervous system via this nerve in 8o%. The remaining 2o% are coming via the sciatic nerve or an accessory obturator nerve. Since the nerve is a mixed one, its motor fibers reach the adductor muscles. The nerve leaves the adductor canal at a point 1 cm lateral and 1 cm caudad to be the pubic tubercle. This point is somewhat more lateral and cranial to the point at which a needle should pierce the skin for a pharmacological block of this nerve [95]. Fig. 71 demonstrates this.

Positions and Sizes of Electrodes: At the point of exit of the nerve from the canal, the anode (7,3 cm^2) should be placed. A moist pad between it and the skin is required in most instances because of the pubic hair. It should be pressed against the skin by a sand bag or preferably a bag containing rice (since this latter does not grow mouldy). To ensure this good skin contact, treatment has to be carried out in the reclining position. The cathode (96,8 cm^2) should be placed over the gluteal muscle opposite the anodal site. The patient should rest on it. The longitudinal axis of this cathode should parallel the course of the obturator nerve.

Indications: (1) Pain in the hip joint; but only, if the joint has not been operated upon since any such procedure, especially for total prosthesis, changes all conditions of innervation entirely. (2) Spasm of adductor muscles. The definite advantage of an electric block of this nerve over a pharmacologic block lies in the fact that (if one will use electric characteristics of a monophasic square wave, 2o-5o times per second, with shorter single impulse duration than 0,2 msec and via a small anode and a very large cathode) only sensory (thin) fibers are blocked and effect on motor (thick) fibers is lacking. This means that the function of the adductor muscles never is suffering from an electric block of the obturator nerve while a pharmacologic block always does interfere with the normal function of the adductor muscles weakening their strength and causing gait disturbance. In most instances, 5 therapeutic session of 20 minutes each suffice to bring about one year of painfree state.

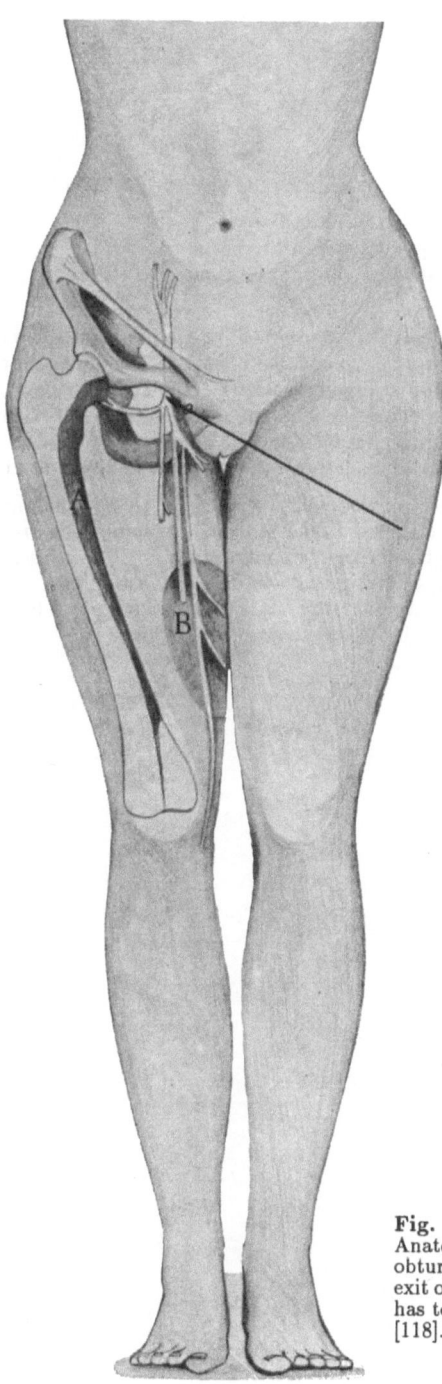

Fig. 71:
Anatomical sketch on position of
obturator nerve, right. Over the
exit of nerve from its canal anode
has to be placed. From JENKNER
[118].

FEMORAL NERVE

Segmental Access: Fibers from the segments L-1 to L-4 contribute to this nerve.

Anatomy: This nerve is to be found lateral to the femoral artery (fig. 72). This latter is situated medial to the mid-point between superior iliac crest and pubic tubercle. 1 cm caudad to the inguinal ligament is the optimal blocking site where the nerve is found about 1,5 cm only under the skin.

Positions and Sizes of Electrodes: A 7,3 cm^2 anode is to be placed 1 cm caudad of the inguinal ligament over the nerve. Attention should be paid to possible hair (in males). Using this position not only the femoral nerve but also the periarterial sympathetic plexus around the femoral artery will be blocked. A 96,8 cm^2 cathode has to be placed exactly opposite to the anodal site. Its longitudinal axis should point parallel to the direction of femoral nerve, i.e. slightly oblique. Electrode positions are marked on the skin in fig. 73.

Indications: Pain having its cause in the supply area of this nerve peripheral to the blocking site. This type of pain is rather seldom. Mainly, the conditions are rheumatoid arthritis, spasms of muscles and metabolic or toxic neuropathies. To have a better understanding of the skin area supplied by this nerve the reader is referred to fig. 74. For causes of pain proximal to the site of blocking but peripheral to a paravertebral block apply the anode as for segmental paravertebral block of the segment involved.

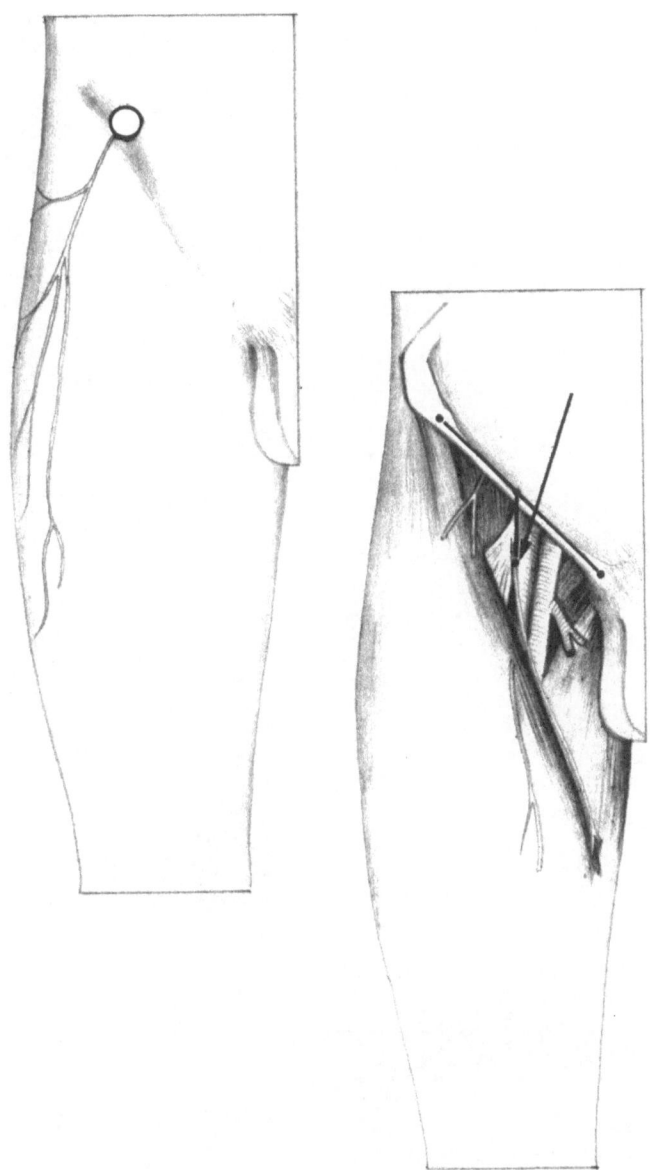

Fig. 72: Anatomic sketch on position of right femoral nerve (right half) and of lateral cutaneous femoral nerve. Sites from which these may be influenced in an optimal way are indicated by arrow and ring (= anodal placements on skin). The femoral nerve is the most lateral of the three structures within the vasculo-nerval lodge of thigh (right: nerve, artery, vein). From JENKNER [118].

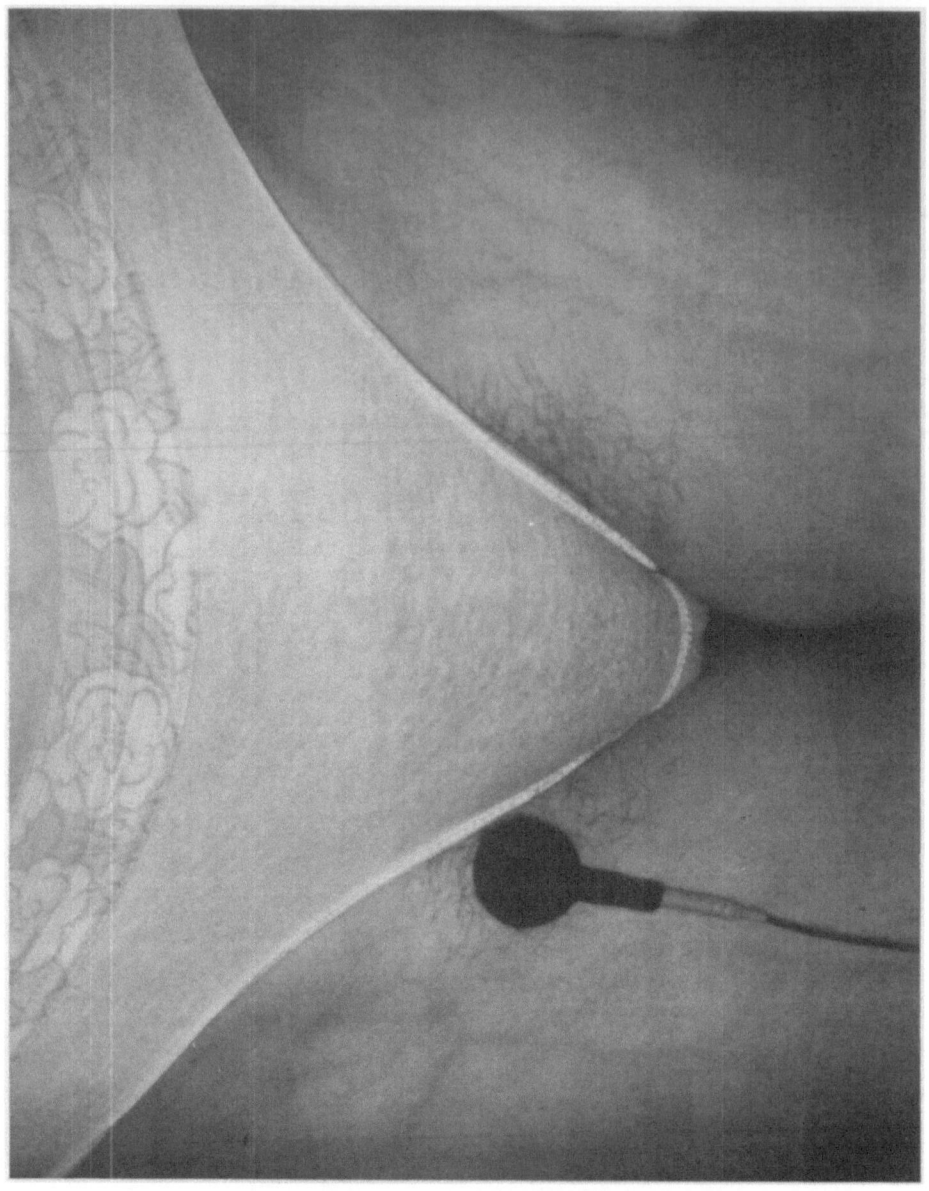

Fig. 73:   Anode in place for blocking pain in the course of right femoral nerve. Cathode ist to be placed over homolateral buttocks.

Fig. 74:   Schematic drawing of skin areas supplied by femoral (densly punctated), lateral cutaneous femoral (dashed) and sciatic (coarsly punctated) nerves. From JENKNER [118]. (facing page)

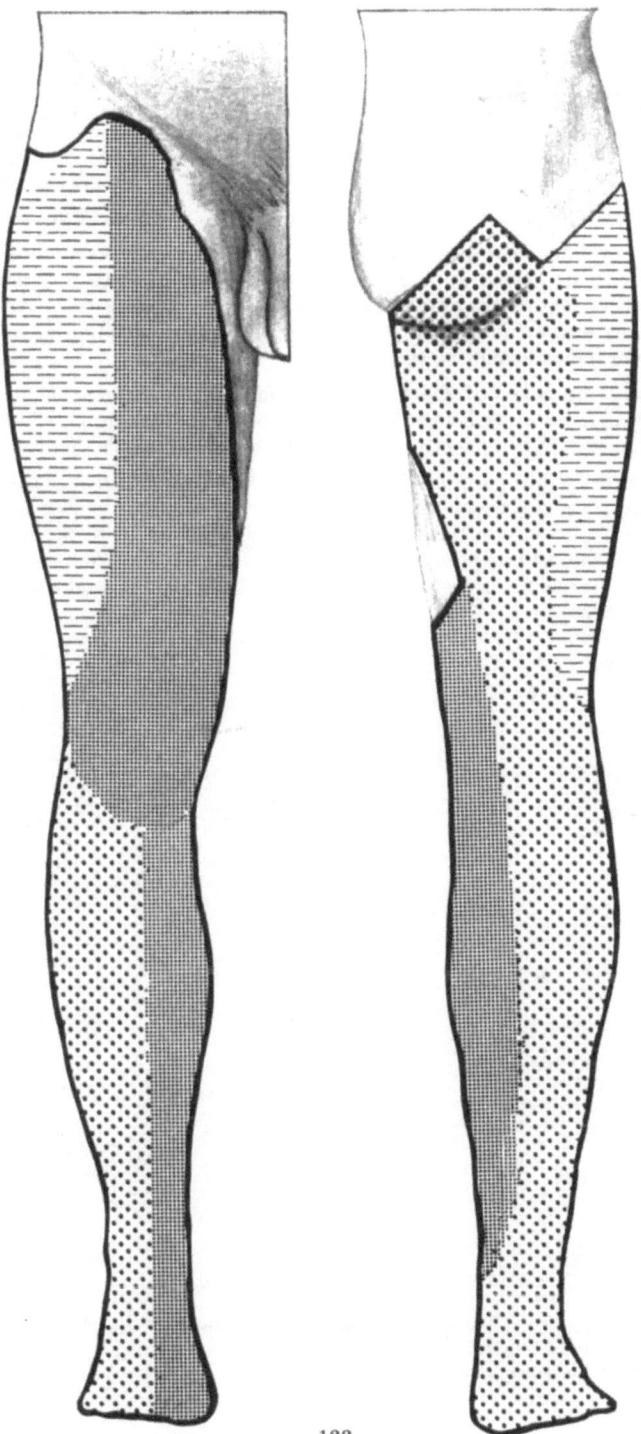

133

LATERAL CUTANEOUS FEMORAL NERVE

Segmental Access: Fibers from the segments L-2 and L-3 form this nerve.

Anatomy: This nerve passes about 2-3 cm medial and just as much caudad of the anterior superior iliac spine (fig.72) to reach the fascia, to pierce it and reach the skin of the area of its sensory representation shown in fig. 74. This corresponds entirely to the painful area where one may also find diminished or lost function of the sweat glands (which however, never should be tested).

Positions and Sizes of Electrodes: An anode of 7,3 cm^2 (or smaller, 1,7 cm^2 also may be used) is fixed using karaya at the point of passage of the nerve mentioned. The 96,8 cm^2 cathode is placed exactly opposite the anodal site on the back of the patient.

Indications: Only pain in the supply area of this nerve, being called paresthetic meralgia. The cause of this condition very often lies in a trauma or damage of some other kind to the nerve, mainly at the site close to the iliac spine. In a correct indication, the results are excellent and may be reached after 5 - 1o sessions of 2o minutes each in daily intervals. Patients usually give clear descriptions of their area of painful sensations making the diagnosis easy.

Fig. 75: Position of anode (upper photograph) and cathode (lower photograph) to electrically block the lateral cutaneous femoral nerve on right.

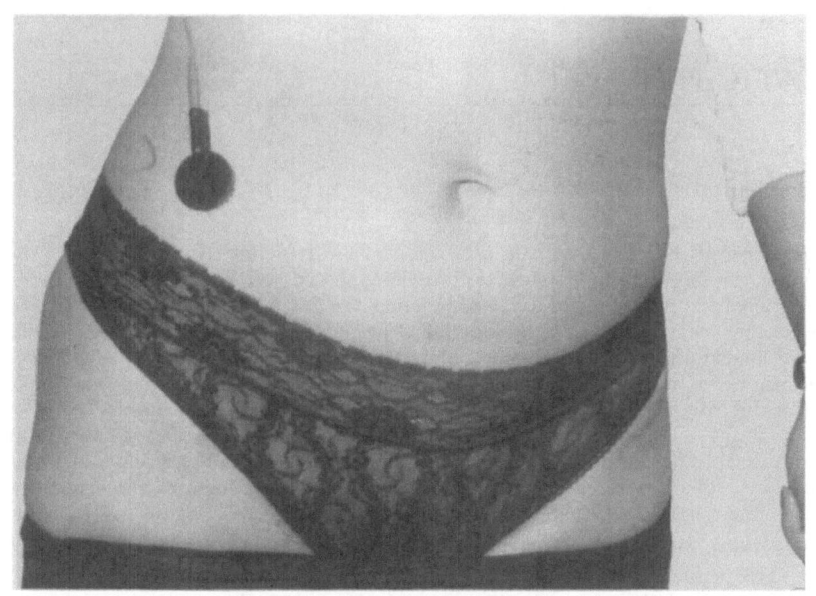

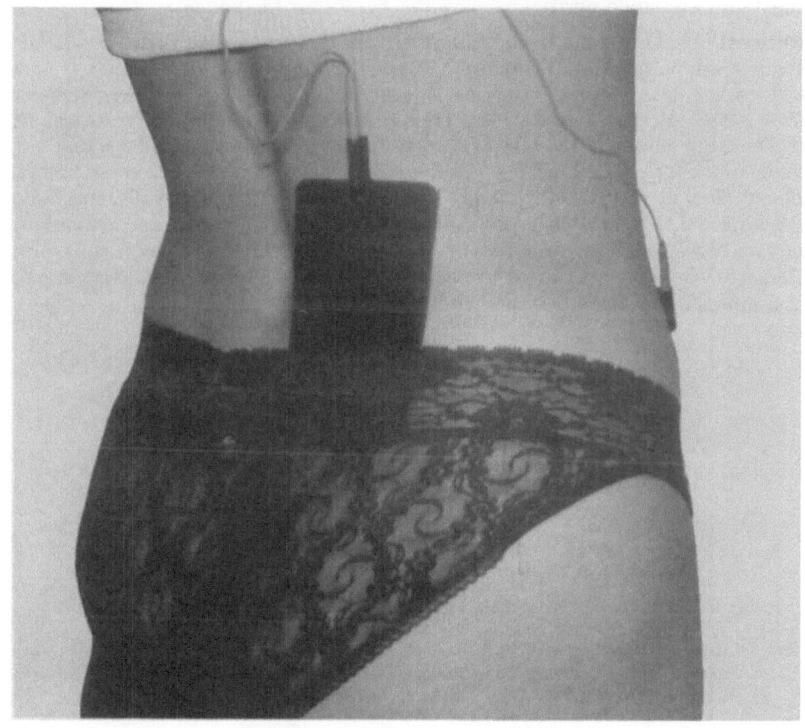

SCIATIC NERVE

Segmental Access: Fibers from the segments L-4 to S-3 are contributing to this nerve.

Anatomy: For blocking this nerve at the most frequently used site it is best to remember the following land marks for reaching the correct position for the anode (fig. 76): Through the mid-point of a line connecting the greater trochanter with the superior posterior iliac spine (LABAT's technique [20] a perpendicular line is drawn and followed in a caudal-medial direction for a distance of 3 cm. At exactly this point the skin is overlying the sciatic nerve as it evolves from under the piriform muscle. - Some 5 or 6 cm more distal of this site, there lies another place at which this nerve may be blocked: Between the caudal and middle third of the distance from the ischial tuberosity and the greater trochanter. Yet another alternative site, but far more distally and therefore less adequate for blocking, since there, one misses the cutaneous posterior femoral nerve to the dorsal aspect of the thigh, is to be found 2 cm caudad of the mid-point of the gluteal fold.

Positions and Sizes of Electrodes: The point to affix the 7,3 cm^2 anode is one of the three named sites. A 96,8 cm^2 cathode is placed exactly opposite on the body surface for each of the three positions named. Its longitudinal axis should be directed parasagittally. Electrode position is shown in fig. 77 for the anode.

Indications: Only pain in the supply area of this nerve with the cause distal to the blocking site will be relieved. Therefore, most of the sciatiforme complaints will have to be treated by other means; which is the case for pain from a herniated disc, fractured vertebra, pain from pathology of vertebral joints. Treatment has to consist of paravertebral blocks of the involved segments as in the case of scars in the segmental distribution of one root. The main indications for sciatic blocking will therefore be all neuropathies (either metabolic or alcoholic or diabetic or toxic) with pain. But these blocks will only be effective if diabetics are well adjusted, external toxic influences (e.g. industrial gases) are avoided and the exposure to these has been terminated. It will also be helpful to electrically block this nerve in cases where pain is caused by spasms of muscles supplied by it. Also in pain from contused muscles this block may be helpful.

Fig. 76:   Anatomic sketch to direct correct placement of anode for electric block of sciatic nerve. Upper drawing: skin mark on bony background to identify landmarks. Lower drawing: skin mark over muscles. From JENKNER [118].

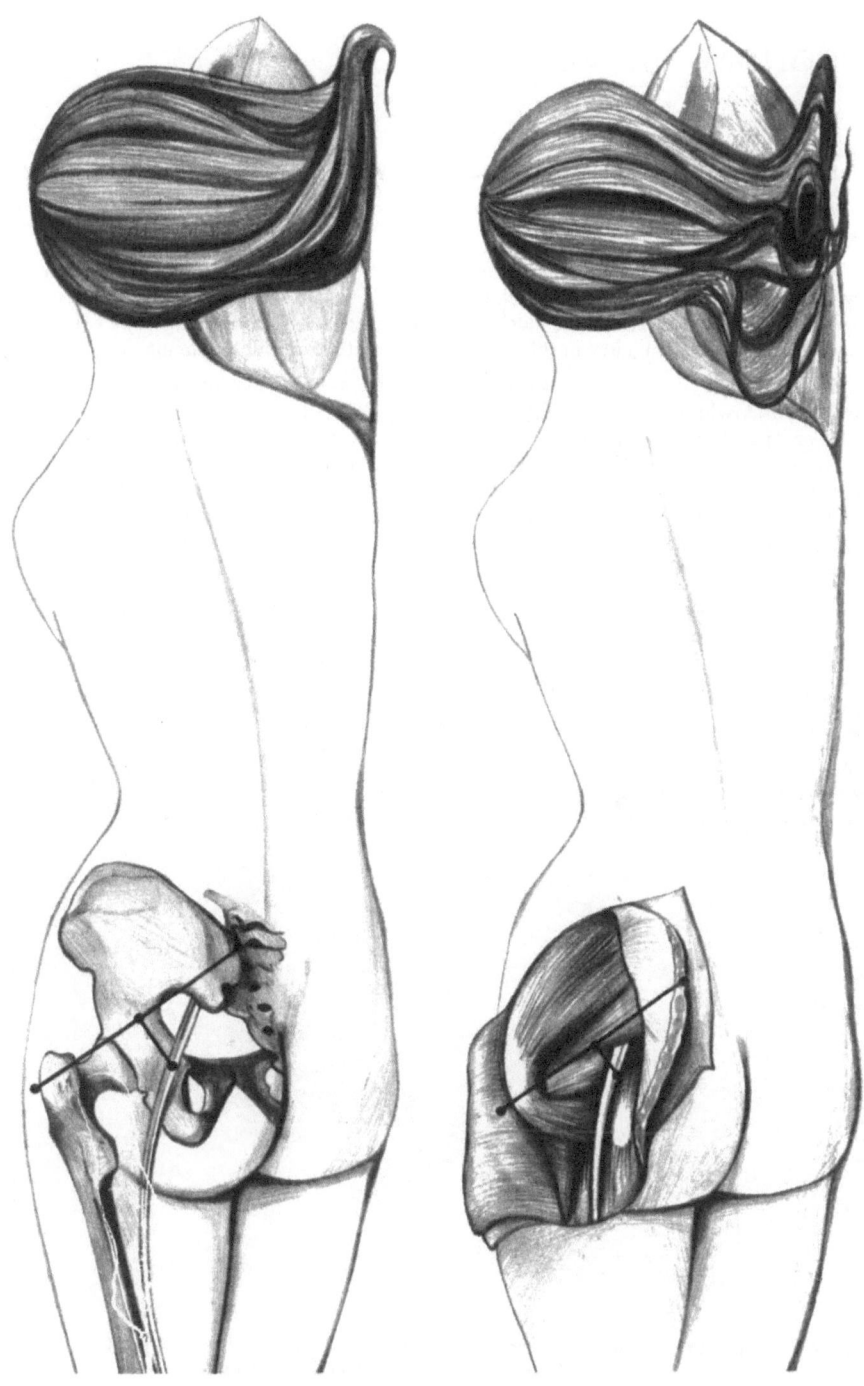

PERONEAL NERVE

Segmental Access: This nerve receives its fibers from the segments L-4 to S-2.

Anatomy: At the site where this nerve passes along the head of the fibula it is best to block it (fig.78).

Positions and Sizes of Electrodes: A 7,3 cm² anode should be placed over the nerve as it passes the head of the fibula (fig. 79). The cathode (96,8 cm²) should be placed at the opposite (which is medial) side of the lower leg and affixed using tape even in case karaya be used since the curving of the skin makes sticking a bit difficult.

Indications: This nerve may not be blocked at the mentioned site in those very frequent cases where the pain originates more proximally, such as is the case in states after disc operations, a vertebral fracture or the like. Only causes of pain peripherally of this blocking site are effectively treated such as metabolic, alcoholic or toxic neuropathies, states after orthopedic procedures resulting in painful states, from scars in the supply area of the peroneal nerve. Also operations of neuromata as well as simple, contusions of the muscles of the pereoneal group, if painful, may be treated effectively. The effect of this electric block usually is excellent because of the very superficial position of the nerve at the blocking site.

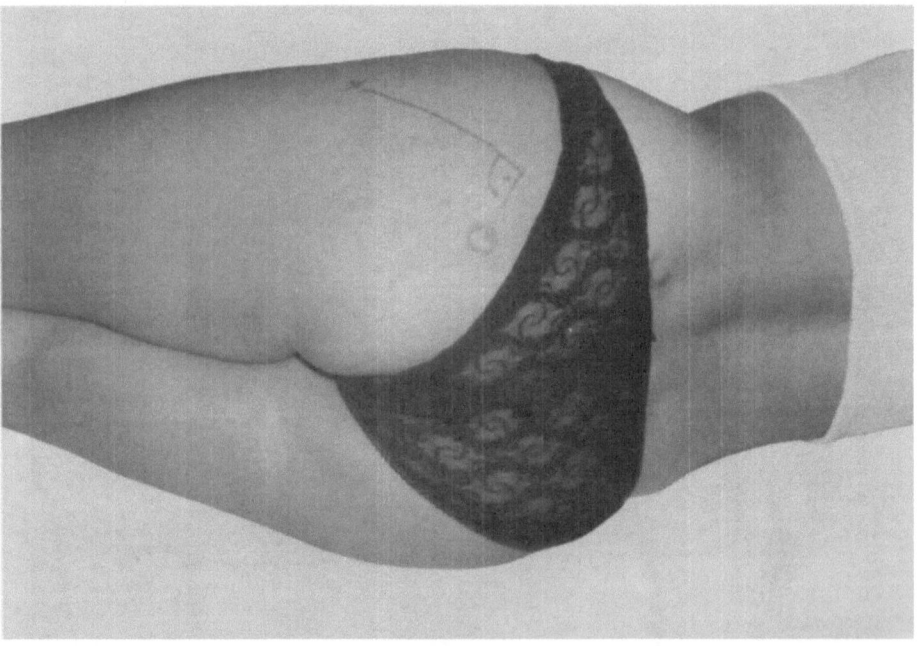

Fig. 77:   Placement of anode for electroblock of left sciatic nerve at skin mark (ring). Cathode would have to be on opposite ventral face of thigh.

138

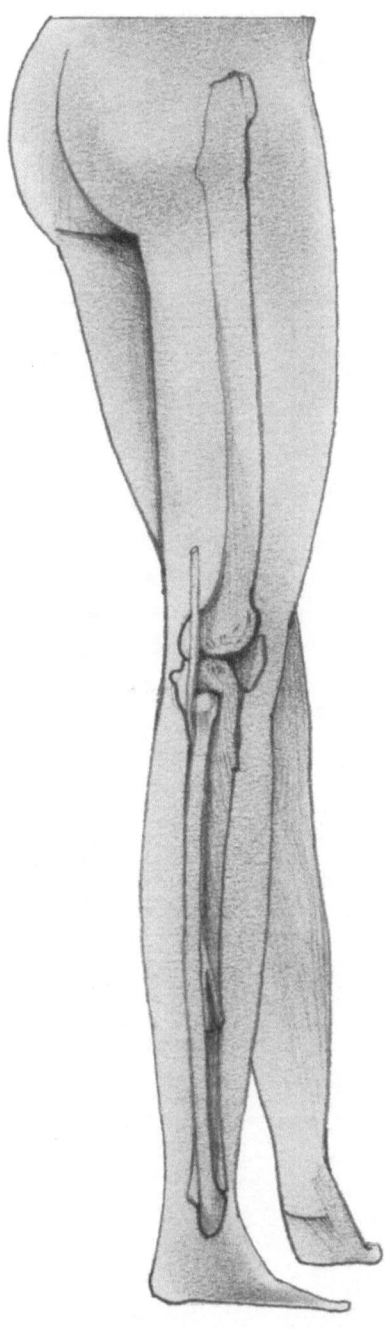

Fig. 78:
Anatomic sketch to help with anode placement for treatment of peroneal pain (right).

139

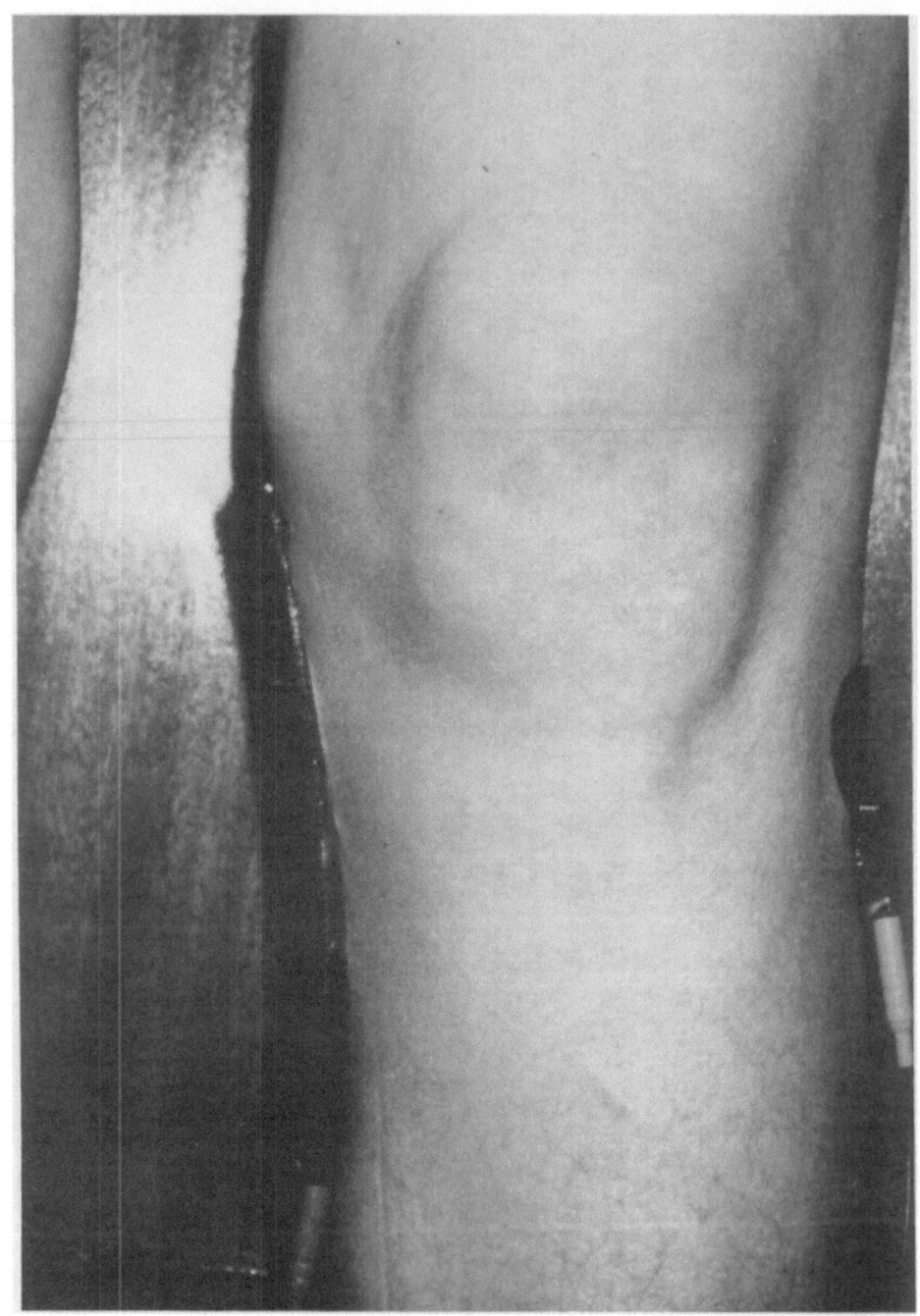

140

140

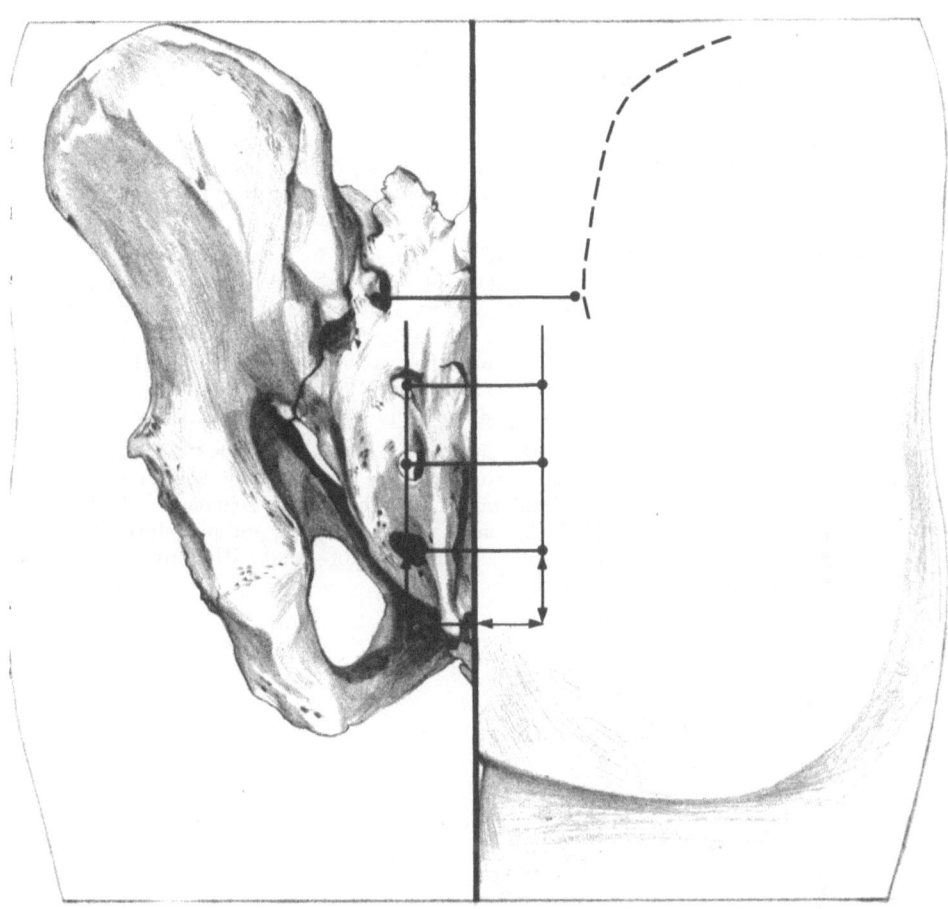

Fig. 80: Anatomic drawing to identify position of anodes for electric block of sacral nerves. From JENKNER [118].

Fig. 79: Electrodes in place for electroblock of left peroneal nerve just posterior to fibular head. (overleaf)

141

SACRAL NERVES

Segmental Access: Supply to these nerves originates from the segments S-1 to S-5.

Anatomy: The sacral nerves pass through the sacral foramina at which site they are situated rather close to the skin and easily within the reach of an electric block: this means effective blocking. The direction of the foramina, which is essential for a block by needle is of no importance for electric blocks. The foramina are easy to palpate usually. This is indicated in fig. 8o (page 141).

Positions and Sizes of Electrodes: An anode of 7,3 cm^2 is to be placed over the respective sacral foramen (fig. 81). A cathode (96.8 cm^2) should be placed homolaterally exactly on the skin surface opposite of the anode. If two (neighboring) nerves are to be blocked on one side, a common cathode on the site mentioned may be used. If identical nerves are to be treated bilaterally (e.g. during delivery) a common cathode in a midline position may be used, but for deliveries two separate cathodes each at a flank ought to be preferred. In case of much hair, a moist pad should be used instead of karaya.

Indications: Pain arising from the segments S-1 of legs, of perineal and perianal regions and the genitalia (labia, vagina, testes) may be treated. Pain from malign causes (e.g. teratoma of testes) may respond differently. In case of a clear segmental origin, about 6o% pain relief may be achieved [146]. Spasm of vesical sphincter and pain during delivery (stage 1) may be treated with good results. In delivery, note, that with porogressing stage different nerves would have to be blocked.

Fig. 81: Position for anode (upper) an cathode (lower photo) to block left first sacral nerve electrically. From JENKNER [118].

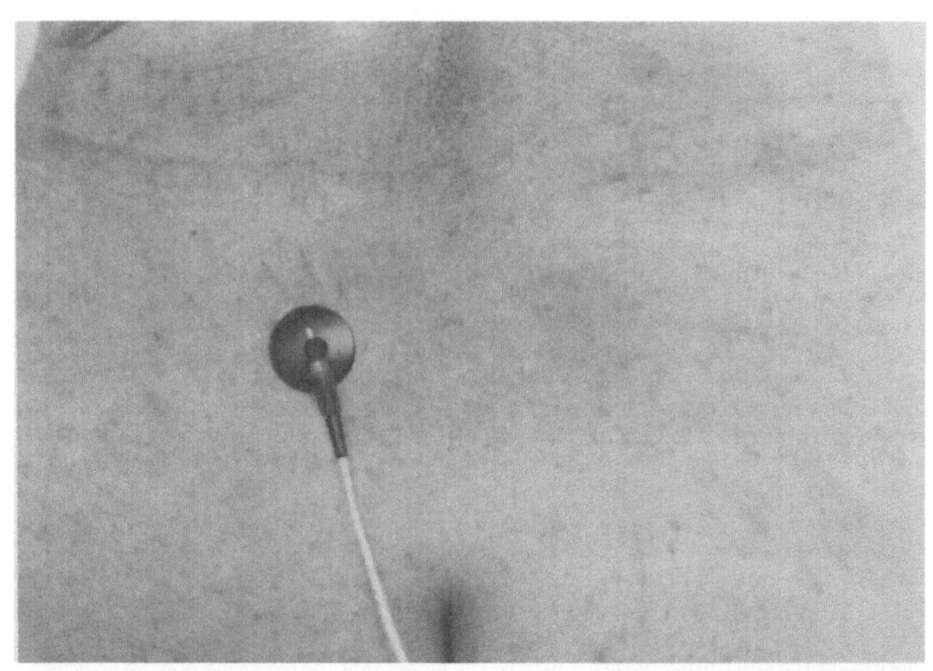

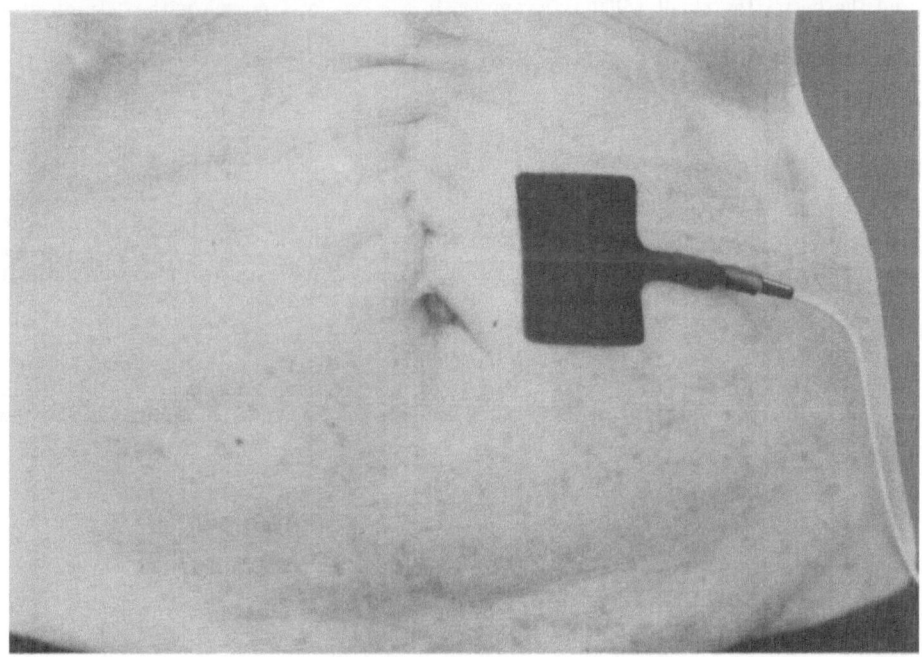

143

PHANTOM PAIN – PAINFUL SCAR

Segmental Access: This depends solely on what "phantom" or scar is giving pain. Segments should be looked up on table 4 (pages 44/5).

Anatomy: It is important to know that phantom pain does not only occur after amputation of limbs or fingers. Every surgical removal of an organ may be followed by phantom pain. Of the number of patients treated for postcholecystectomy syndrome not just a few are actually suffering from phantom pain. Removal of a breast, total gastrectomy, even an appendectomy may be followed by phantom pain relating to the removed organ. One should think of this possibility to recognize it. Rather frequently this occurs after the enucleation of an eye ball. There also exists a so-called "visceral phantom pain", which occurs after vagotomy for peptic ulcer: inspite roentgenologic evidence of healing of the lesion ulcer pain (and other symptoms) persist. Treatment should then take place by electric nerve blocking of the somatic nerve responsible for pain conduction.

Every scar may give rise to a neuralgiform pain. This then is due to either surgical sectioning of the ever so small (and therefore invisible) branch of a nerve or by tying this invisible nerve into a suture.

Positions and Sizes of Electrodes: The anode (usually 7,3 cm^2) should be placed over the nerve responsible for pain transmission or at the painful spot at the scar. The more exact the description of such a pain is, the more exact the electrode may be positioned and the better the result will be. Example: In the case of a person with a stump of a thigh, the painful area is the foot with emphasis on the great toe (even though amputated: this is characteristic of true phantom pain): here the optimal placement would be over the root L-5 ipsilateral or at the cut end of the sciatic nerve. The only exception to this would be, if the amputation had been carried out for vascular reasons: then the anode needs to be placed over the lumbar sympathetic at L-3 (to L-5). In all instances the cathodes should be 96,8 cm^2 and be placed on the opposite surface of the body (or stump). In the case of a painful scar after an appendectomy (fig.82) place the anode over the painful site of the scar or proximal on the course of the respective intercostal nerve and the cathode on the back ipsilaterally opposite the anode. Therapy should be 2o minutes every day for one month, sometimes up to three months are required in patients with phantom pain. If daily treatments are carried out, satisfactory results may be expected (fig. 82a).

Indications: Phantom pain, but also stump pain and painful scars may be treated. One calls a pain, where the stump and not the removed area of a limb is painful stump pain. With stump pain 2 - 3 weeks of daily treatment sessions are sufficient. But also in true phantom pain, this type of therapy requires correct placement of the anode over the nerve responsible for pain conduction.

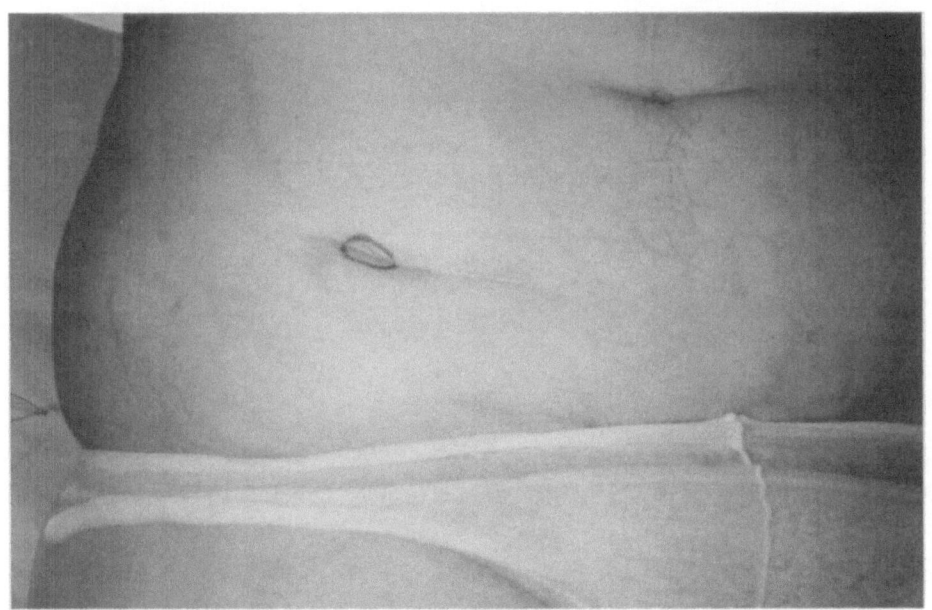

Fig. 82: Placement of anode for treating painful scar (after appendectomy, as in this case) electrically. The cathode is to be placed homolaterally on the skin of the back, opposite of anode, its long axis parallel to the main direction of the scar.

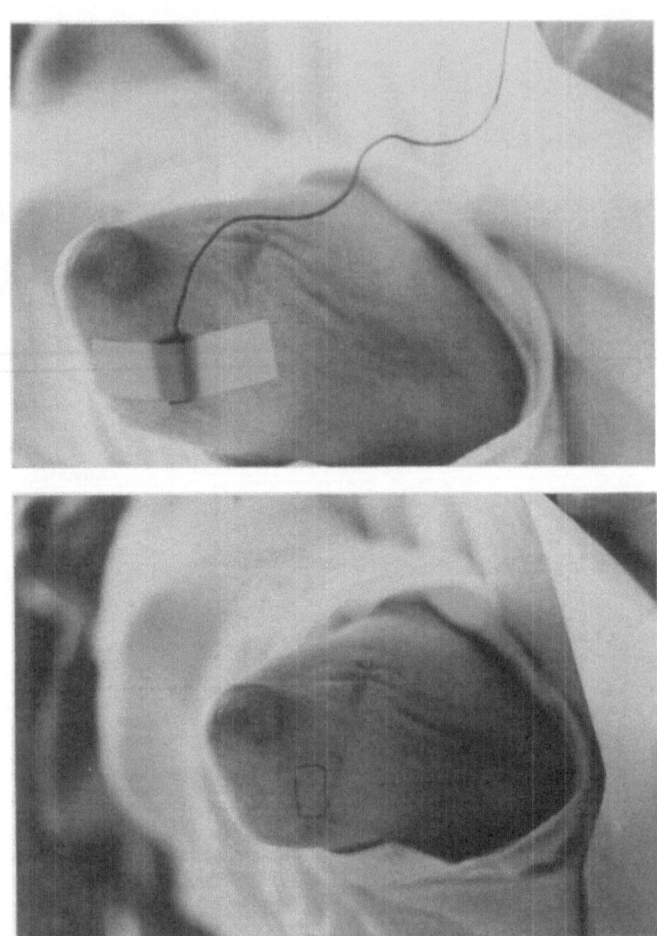

Fig. 82a: Position of anode in a case of phantom pain: left figure shows skin mark, right figure shows anode in place. From JENKNER [118].

TRIGGER POINTS AND MYOFASCIAL PAIN

The importance of trigger points and their recognition has been emphasized by many. The latest survey by TRAVELL and SIMONS [236] is a comprehensive summary and provides the reader with many useful drawings. DE MAR [52] has combined the typical electrical impulse described on page 21 for nerve blocking with the therapy of pain via trigger points (by other means). He suggests placing the small anode over an established trigger point after having arrived at a clear cut diagnosis. The cathode is placed on the surface of the body opposite of the anodal site. Following his suggestion we have used this method and found it to be valuable in cases of nodular trigger points within the supraspinatus or levator scapulae muscles. These applications DE MAR already mentions and we could substantiate the positive effect anodal placements over these trigger points have on pain irradiating from these muscular lesions. Since we are unable to recognize any mechanism of action of influences on trigger points the assumption remains unproven that monophasic impulses are influencing small nerve fiber endings in the trigger point area to diminish their conductive capacity resulting in reduced pain sensation. The only empirical application of the type of electrical impulses mentioned suggests limitation to the therapy of myofascial pain.

JOINT PAIN; NON-PAIN-ASSOCIATED STATES

A clearly defined form of electrical impulses may be determined for over-active or irritated states of the sympathetic nervous system having a blocking action on these nerves. While this same kind of electrical impulses is optimally suited for electric pain control, no such statement is possible for those states where pain is situated in joints. This is mainly due to the difficulty to name a certain nerve being solely responsible for pain conduction. Experience has shown that in these cases a somewhat higher content of DC is needed. What empirically has been seen is difficult to realize because not many sets commercially available will provide such kinds of impulses. Most sets however provide the possibility of increasing the duration of a single impulse by using a very small screw driver through a small hole in the case of the set. It is advised to double the duration of single impulses if no proper impulse having a higher DC-component is provided by the instrument. This doubling relates to the original duration of a single impulse as is customary for use with small diameter fibers (pain conduction and sympathetic nerves). It is best to use instruments where the patient has no possibility to change any characteristics but the intensity of the current. We also prefer instruments supplying the cable connection via a coaxial cable to the electrodes thereby preventing wrong polarity being used by the patient. This would require however a different coaxial cable for other applications than for pain and sympathetic dystrophies, as e.g. for motor effects.

In treating painful joints (and ligaments) the optimal size of an anode is 7,3 cm² and that of a cathode 96,8 cm². The anode preferably should be placed over the most painful area. The frequency of impulses is identical with that use for pain. In this section examples for application on painful joints will be found under the numbers 1 - 8.

Examples for none-pain-uses will be numbered 9 to 12. For these the characteristics of a single impulse are identical with these given for treating painful states or overactive sympathetic nerves, with the exception of two items: for achieving motor effects or trying to influence a muscle directly, the frequency should be 1oo/sec or higher (while 35/sec will work too, but not as efficiently) and polarity must be reversed, meaning a very small cathode and a large anode should be used with the small cathode over the nerve (or on the muscle) one wants to influence. This has to be followed because the irritating or stimulating effect prevailing under the cathode (fig. 86) should be maximal to stimulate a motor nerve or a muscle. The soothing effect under the anode (fig. 86) desired for nerve blocking in case of pain has to be minimized in these instances. Example for a motor effect is listed as number 10 (and another has been seen in the application of an electric impulse to the phrenic nerve; see page 103). Applications to muscles directly are described under the numbers 9 and 12. Therapy for the vagus nerve (number 11) should also be done using a frequency of 35 and the polarity as for treating painful states. As such, there are enough non- pain-associated states (JENKNER, [115]) for which the named form of impulses may be used. We are not going too much into any details here leaving it to the reader to select those among his patients for whom he/she may be able to use this principle advantageously. To stay with a lower frequency in cases of desired motor effects will remain for those instances where an instrument has no possibility to change its frequency to a higher than 35/sec value.

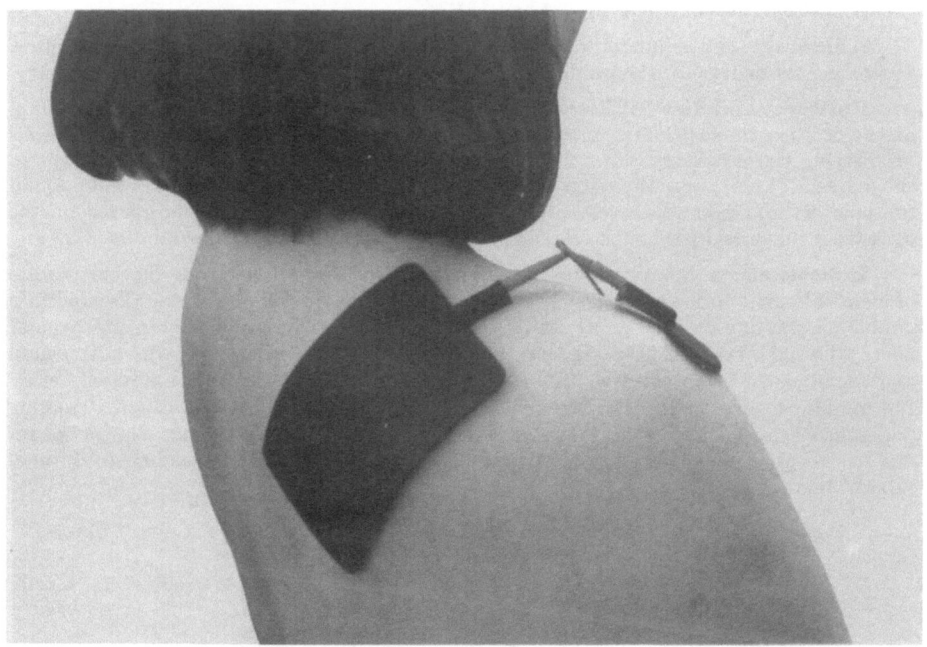

Fig. 83: Positions of anode and cathode for electrically treating periarthropathic pain locally (of right shoulder) using brief DC-impulses.

1. HUMERO-SCAPULAR PERIARTHROPATHY

Segmental Access: The shoulder joint is supplied by fibers from the fifth, sixth and seventh cervical nerve, rarely also in part from the fourth nerve. To reach their destination fibers run through the subclavian, sub- and suprascapular nerves, occasionally also the thoracic and the axillary nerves. Should the patient be able to give a precise description of the distribution of pain, one may be able to differentiate between a segmental or a nervous distribution of it (see fig. 14 and 15).

Anatomy: The peripheral nerves reaching the joint are best met by electric impulses at sites mentioned under the heading of the respective nerve.

Positions and Sizes of Electrodes: Placing the electrodes differs widely depending on the decision to treat locally or by affecting one or the other nerve. Electrode positions for treating via nerves are shown under the respective nerves. For local treatments, place the anode (7,3 cm^2) over the anterior border of the acromion, where there is to be found the point of maximum pain very often. The cathode (96,8 cm^2) is to be placed on the back from the dorsal part of the shoulder to the spina scapulae, as shown in fig. 83.

Indications: In humero-scapular periarthropathy several electrode placements are of identical value and are equally effective in reducing pain. This has been observed in a double blind study of AMMER, K. and F.L.JENKNER [5]. Therapy via the suprascapular nerve, the sixth cervical nerve and local electrotherapy were compared with ultrasound and found much more effective. For omarthroses only local therapy is advised, using the impulse having higher DC- content or a longer single impulse duration. Treating periarthropathy, for local therapy also use these impulses while for therapy via the nerves one has to use the regular very short impulses described under the heading of optimal form of therapy (pg. 21).

2. COSTEN SYNDROME (T.M.J.-PAIN)

Segmental Access: Runs via fibers of the trigeminal nerve. Up to 4% of the patients said to have this neuralgia may have complaints actually due to this syndrome and not to trigeminal neuralgia.

Anatomy: A normal occlusion of teeth secures a normal play of the muscles essential for chewing with no disturbance of the congruent faces of the jaw joint.If occlusion is disturbed such as may be the case in poor prosthetic dentures or non-optimal dental repair of a conservative nature or by loss of teeth and poor dental hygiene the joint faces will be abnormally strained and pain ensues. This pain usually is unilaterally and located pre-auriculary. The jaw joint may be painful to pressure mainly unilateral.

Positions and Sizes of Electrodes: In unilateral pain, an anode should be placed directly over the joint and have a size of 7,3 cm^2. A cathode of identical size should be placed over the contralateral, non-painful joint. Impulses having a higher DC-component should preferably be used. In bilateral pain, switch the positions of anode and cathode for the daily treatment sessions daily from one side to the other.

Indications: Only if there really exists a COSTEN syndrome this position of electrodes should be used. Sometimes pain may radiate to upper jaw, temple and ear. Pain may be constant and exacerbate during chewing and/or talking. This therapy will only be of definite help if measures are taken simultaneously to ascertain a correct occlusion. Up to the establishement of such conditions, daily treatments of 2o minutes each should be helpful in diminishing the pain of this syndrome.

Remark: T.M.J.-Pain is much more common in persons living in the USA., since most people are chewing gum there, thereby overstraining the temporo-mandibular joint their life long. It was found that for this syndrome (or in general for pain from tired muscles or tension headaches) a special type of current seems to be of help. This current consists of very slow pulses (0,66 per second) with a somewhat longer puls width (300 μsec) and a biphasic shape (Attention: special case!). It has been shown by DICKEL et al. [54] that this type of impulse is able to fill up the depleted stores of phosphates within the muscle tissue and is reducing pain if caused by muscle exertion (DICKEL, [55]). These ultra-low frequencies (LANGENBERG, [156]) stimulate endorphine production and lead to relaxation of muscles. Therefore, this form of current should be used for the named indications.

3. RADIAL AND ULNAR EPICONDYLITIS

Segmental Access: Local nerve endings are derived from the seventh cervical root. This fact remains essential if a true epicondylitic process does not exist and one week of therapy is not helpful. Then projected pain must be supposed to exist and this nerve will have to be selected for effective treatment. See fig. 14 and 45.

Anatomy: It should be stated that all modern methods of surgical approach to this pain are directed towards denervating the painful structures (epicondylus and tendons inserting there). To this same end electric blocking of these structures may be used. Whether caused locally or by pain projection this electric therapy will be very effective in reducing pain.

Positions and Sizes of Electrodes: An electrode of 7,3 cm^2 (a very small 1,7 cm^2 electrode also may be used) directly over the involved epicondylus. The large cathode (96,8 cm^2) on the opposite surface of the elbow (fig. 84). Use impulses having a higher DC component.

Indications: True epicondylitis and projected pain at these same sites may well be treated locally using the electrode position named. If after one week no convincing effect is observed one should then switch to treating the seventh cervical nerve.

Fig. 84:   Small anode (upper photograph) and larger cathode (lower photograph) in place for treating right ulnar epicondylar pain.

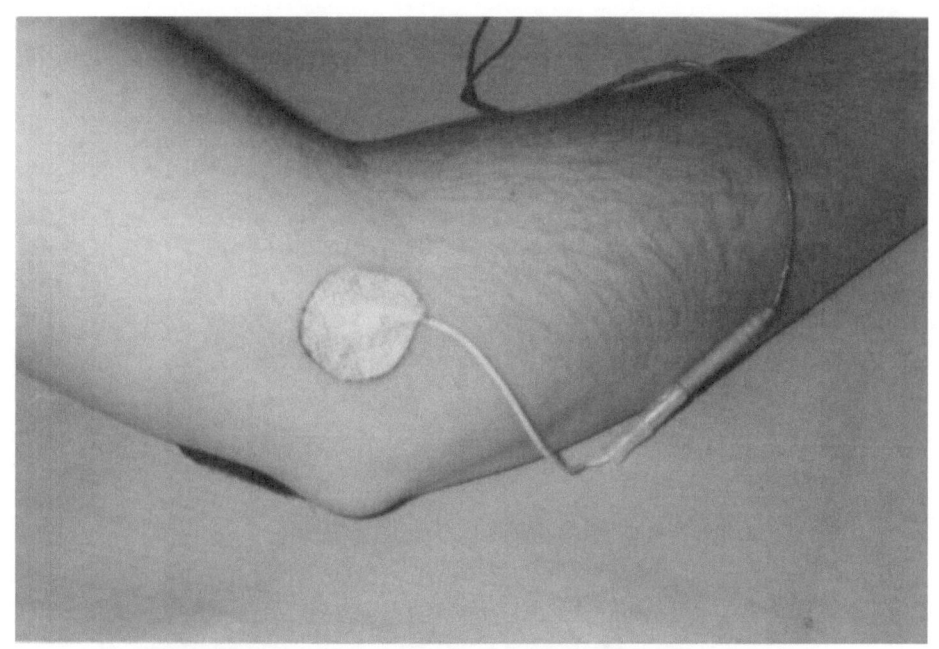

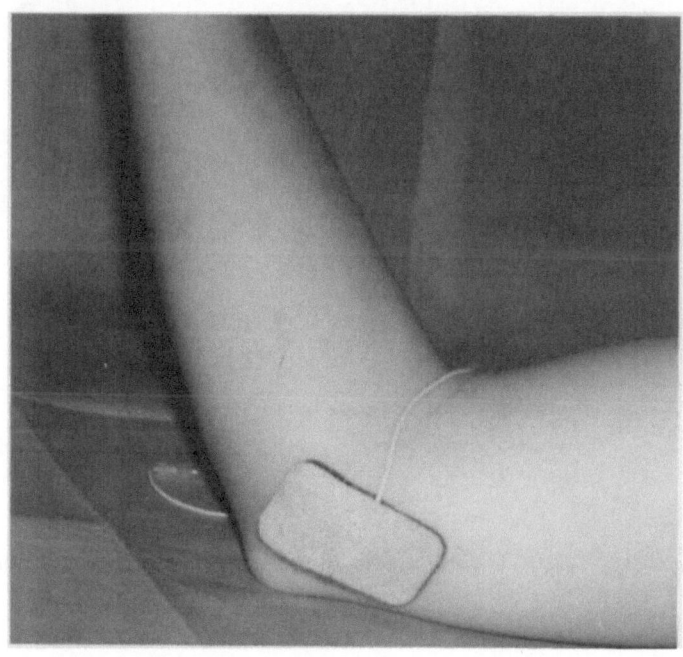

153

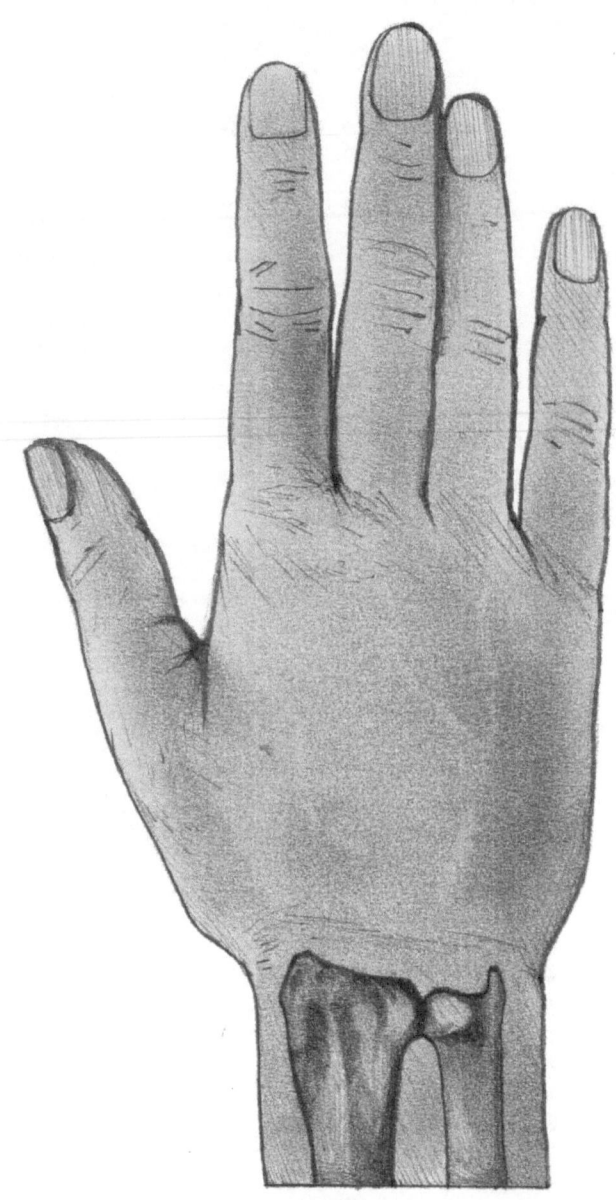

Fig. 85:   Anatomic sketch to identify styloid processes of radius and ulna.

154

4. RADIAL AND ULNAR STYLOIDITIS

Segmental Access: Fibers to these processes run via the radial (or ulnar) nerve or the sixth (eigth) root respectively.

Anatomy: The styloid processes are depicted in fig. 85. There is no special advice except if local therapy fails the nerves projecting to these sites should be recalled.

Positions and Sizes of Electrodes: An anode (1,7 cm^2) should be placed over the involved styloid process. A cathode (7,3 cm^2) has to be placed to the skin surface opposite the site of the anode at the wrist. See fig. 86. Type of impulse as for joints (increased DC-component).

Indications: A first therapeutic attempt should be made locally. If after a week no very convincing result is obtained switch to the respective cervical nerve.

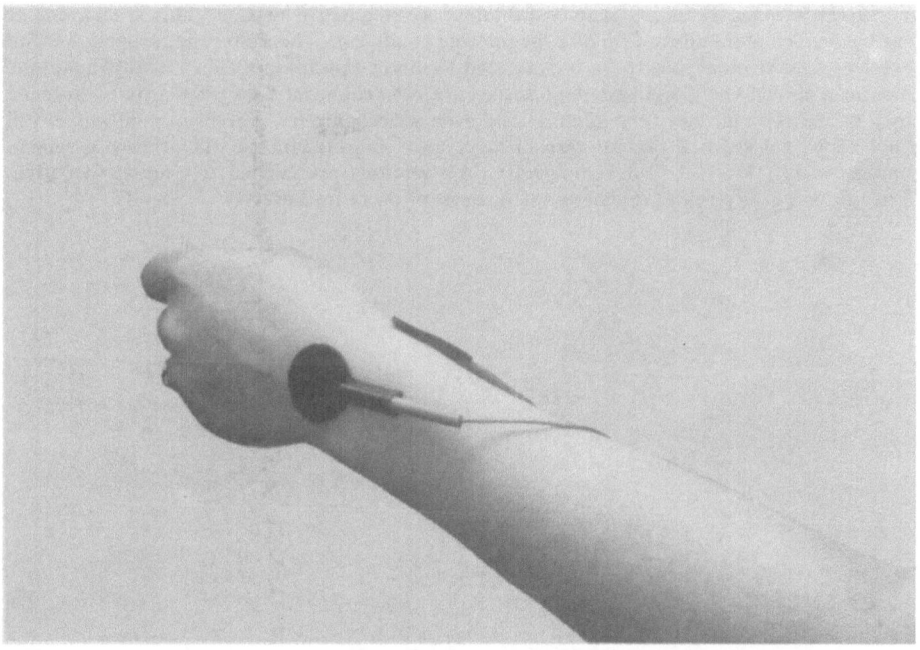

Fig. 86: Position of anode and cathode (met tangentially and barely visible) to treat electrically pain from styloiditis of radius on left.

155

5. STERNO-COSTAL AND COSTO-CHONDRAL PAIN

Segmental Access: The costo-chondral junctions as well as the sterno-costal joints are supplied by the respective intercostal (somatic) nerves running on the lower groove of the rib in question. In both instances, the anterior branch is involved.

Anatomy: No special remarks required. A sketch of the respective areas is presented in fig. 87 for the sterno-costal area and in fig. 88 for the costo-chondral junctions.

Positions and Sizes of Electrodes: A 7,3 cm² anode should be placed directly over the painful sterno-costal joint (fig. 89a) and the cathode (96,8 cm²) directly opposite the anodal site on the patient's back. Both types of impulses, the one without DC-component as for pain therapy and the one with higher DC content (or double the duration of a single impulse) as for joints may be used. A combination of the two applying each type of impulse for 1o minutes each instead of one type of impulse for 20 minutes daily may be tried and should be equally effective. Identical directions may be given for treating costo-chondral pain for which electrode positions are shown in fig. 89b or treating sterno-clavicular pain (figs. 90a + b).

Indications: Painful sterno-costal joints are frequently swollen. This is regarded as early evidence of manifestation of a rheumatoid condition. Therefore, the respective blood cytology and serology has to be investigated to direct therapeutic effort to antirheumatic measures should the blood tests be positive. – Costo-chondral pain frequently is preceded by a trauma (sometimes very minimal and even unrecognized). The proper questions will have to be asked for a precise case history, since microtraumata will otherwise remain unrecognized. We feel that radiological investigations are carried out much too often. They do not really reveal much except in cases of overt fractures.

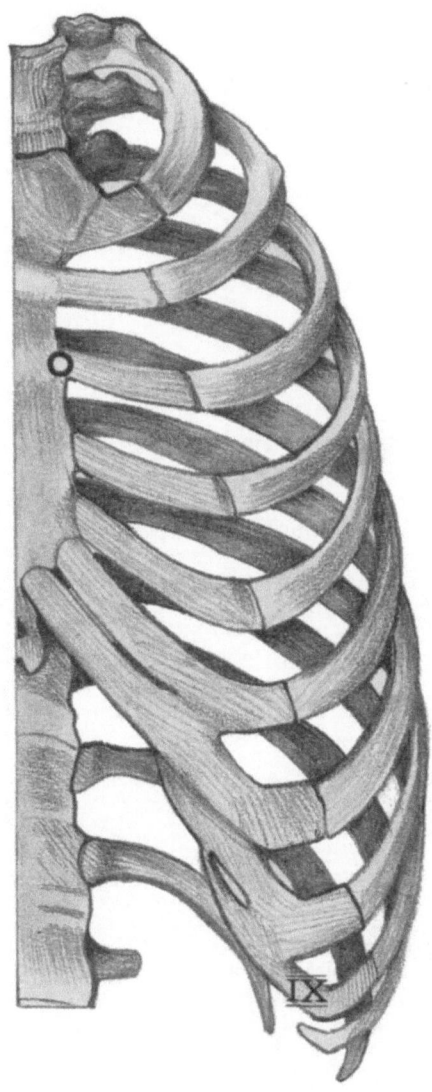

Fig. 87: Anatomic sketch showing sterno-costal region on left side. This should help to place anode for treating sterno-costal pain locally.

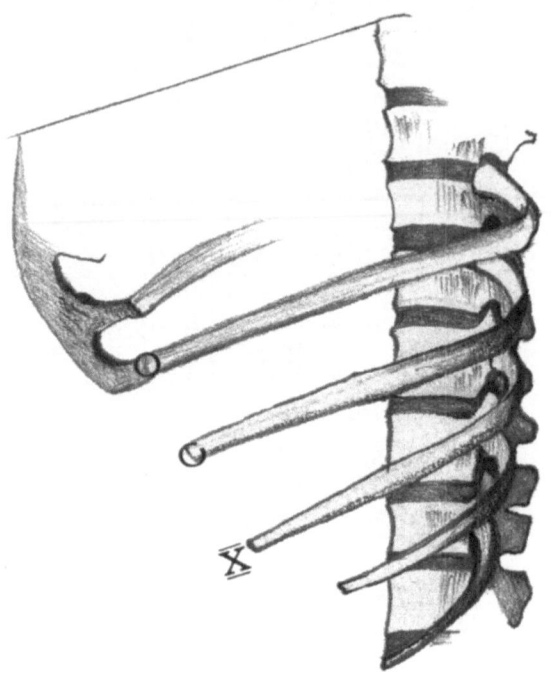

Fig. 88 :   Anatomic sketch of costo-chondral region.

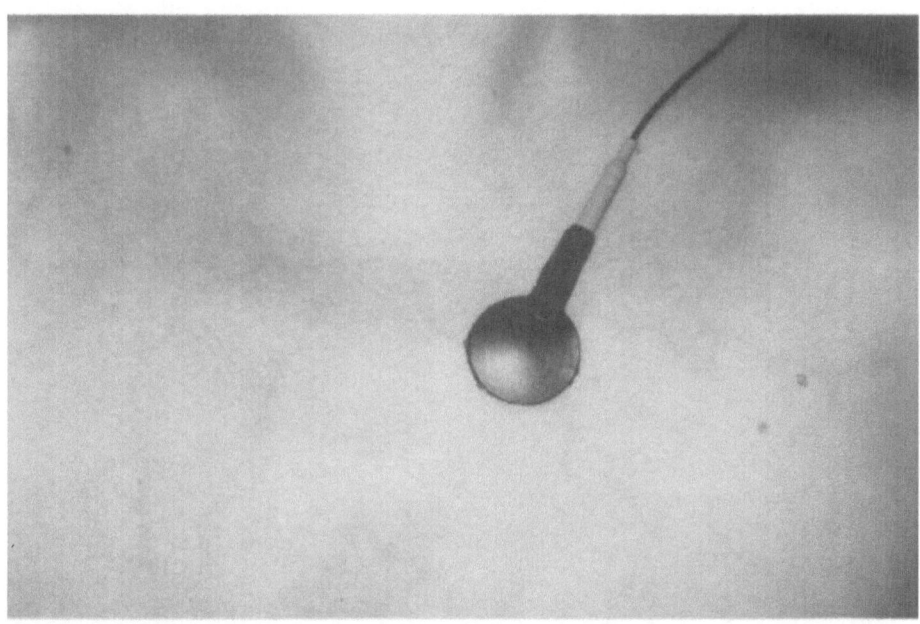

Fig. 89a: Electrode position for electric treatment of sterno-costal pain at second sterno-costal joint. Position of anode shown. Large cathode is to be placed on skin surface of back opposite of anode.

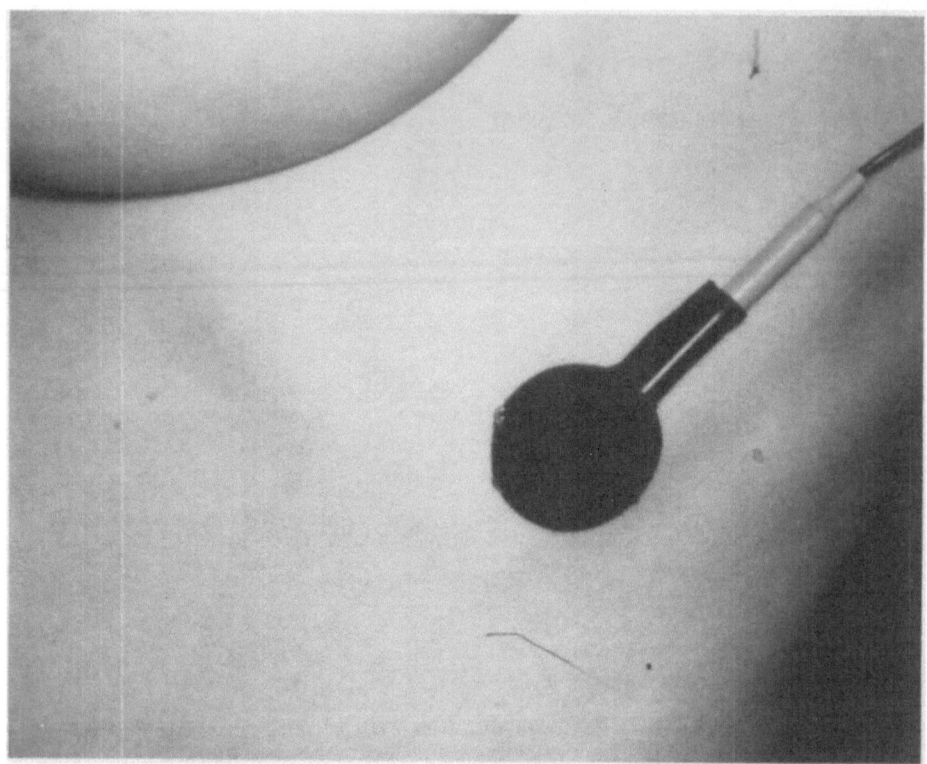

Fig. 89b: Position of anode for the treatment of painful costochondral junction (ninth rib on left side) shown. Large cathode to be placd on same side of back, opposite of anode with long axis of same in the direction of intercostal nerve under anode.

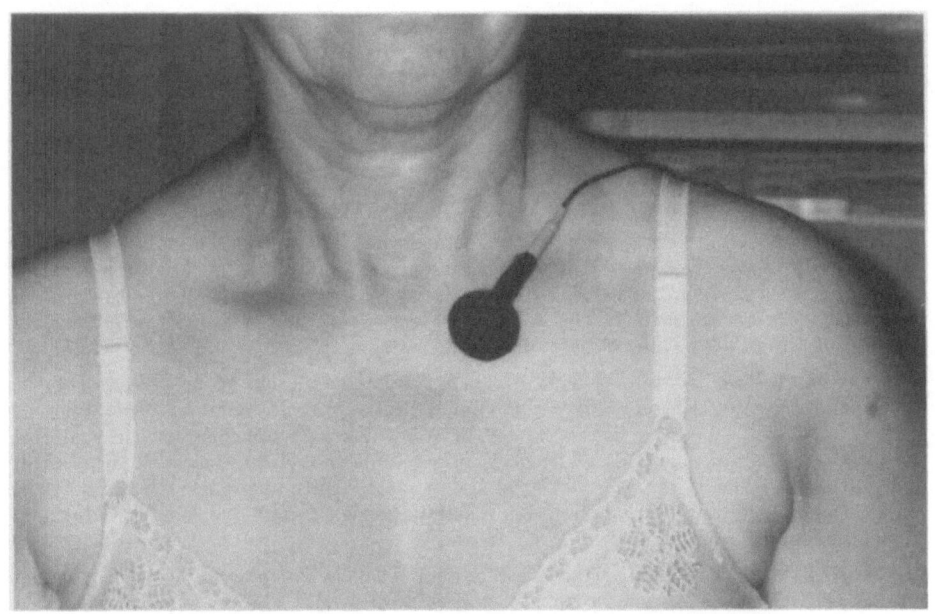

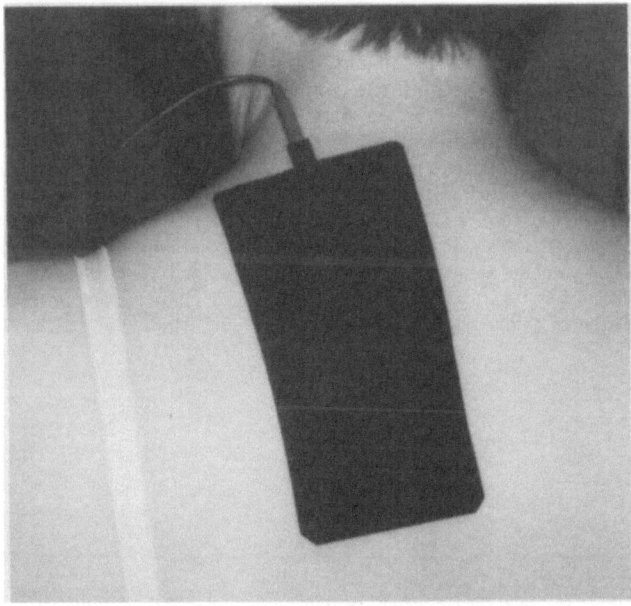

Fig. 90 : Electrode position (a) for anode and (b) cathode to treat sterno-clavicular pain in a respective patient. Note swollen articular area on left.

6. SACRO-ILIAC PAIN

Segmental Access: The sacro-iliac joint is relatively long, situated obliquely and supplied by fibers from the segments L-3 to S-1, occasionally also by S-2.

Anatomy: Only very minimal movements are possible in this joint. The position of the joint with its very uneven articular faces is seen in fig. 8o. There are existing a number of distortions of normal ossification around this joint resulting in several small separated bones at the normal course of this joint. These may only be revealed by radiological studies but are of little value in case of painful states.

Positions and Sizes of Electrodes: A 7,3 cm² anode has to be placed over the most painful part of the joint. Since this joint is relatively large, this may be at different parts of the joint. In most instances, the most painful area coincides with the dimple visible on the skin of the back. The 96,8 cm² cathode should be placed on the ventral face of the body directy opposite of the anodal site. Its longitudinal axis should follow the oblique axis of the joint (or its slit). For a relatively high (up on the joint) painful area the electrode positions are shown in fig. 91 and 92.

Indications: Sacro-iliac pain is a relatively frequent complaint. Often it is being misnamed as "sciatica" or simply "low back pain" which term so often is (being) misused to mean disc disease. Actually the involvement of the sciatic nerve is a "pseudo-sciatica" as the proximity to the joint sometimes causes progression of inflammatory disease of the joint towards the nerve (or its roots). 16% of the painful joint condition in our patients had been caused by some kind of infection. It is advisable to check the blood status including rheuma serology. Even subacute infectious processes of the joint will prevent full effect of any other than anti-infectious measures in treating this painful condition. If no inflammatory process is causing the pain, 5 daily treatments with impulses having a somewhat higher DC-component will suffice to bring about painfree state to the patient. – Of 1500 patients referred to our pain clinic as having low back pain or sciatica 40% actually had sacro-iliac pain and our therapy brought full relief. We advised our patients to add a special active exercise (after ALEC THOMPSON) every evening in a relaxed supine position. Dr. THOMPSON's self-help sacro-iliac replacement maneuver is done by pulling one knee up and laterally at an angle of 45° using the ipsilateral hand and seesawing five times, the contralateral hand is steadying the heel as close to the perineum as possible meanwhile. These instructions are to be repeated with the other knee always exercising both sides.

Fig. 91: Anode in place for treatment of pain in right sacro-iliac joint (to page 163).
Fig. 92: Cathode in place for treating right sacro-iliac pain (lower photograph).

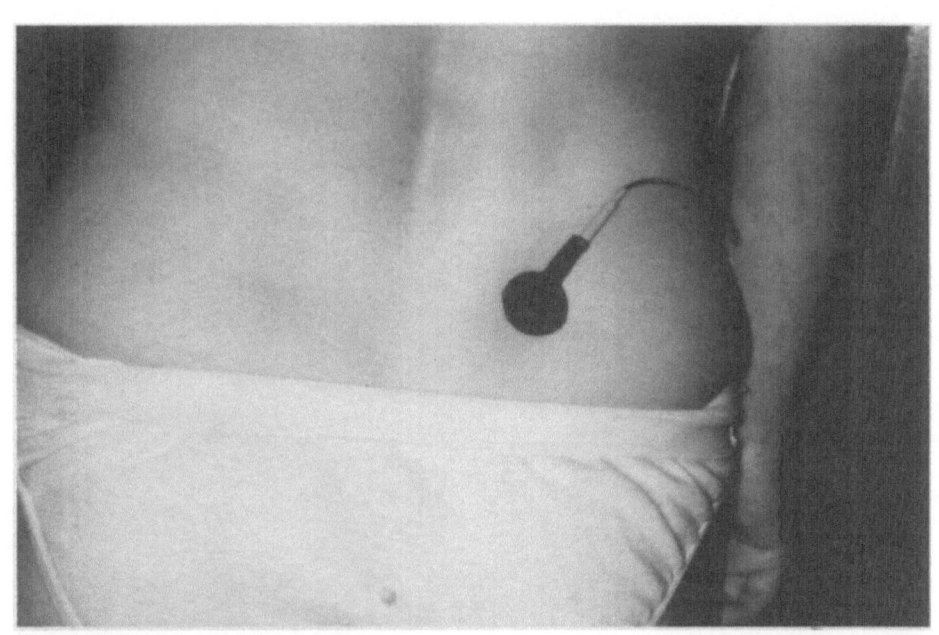

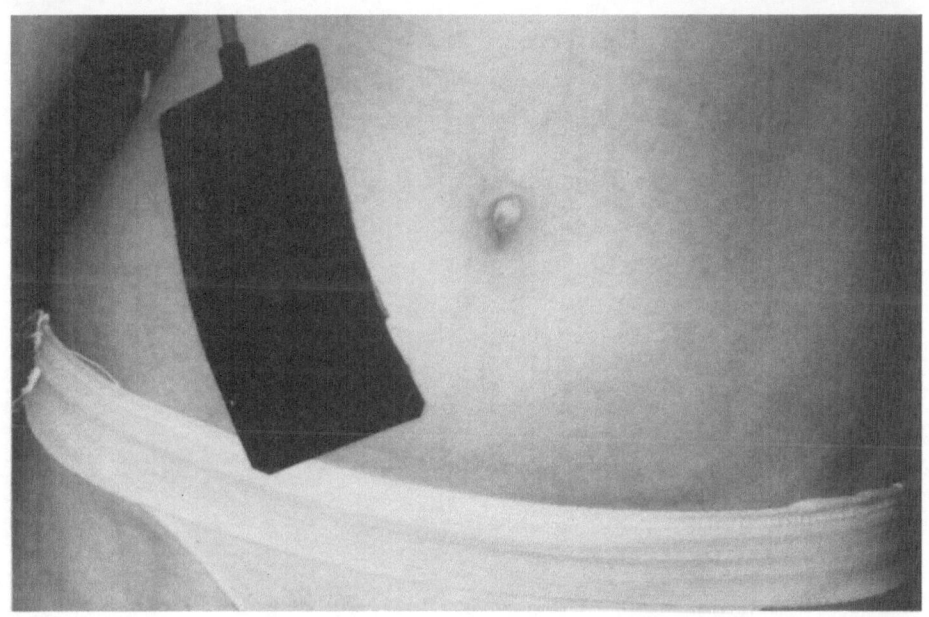

163

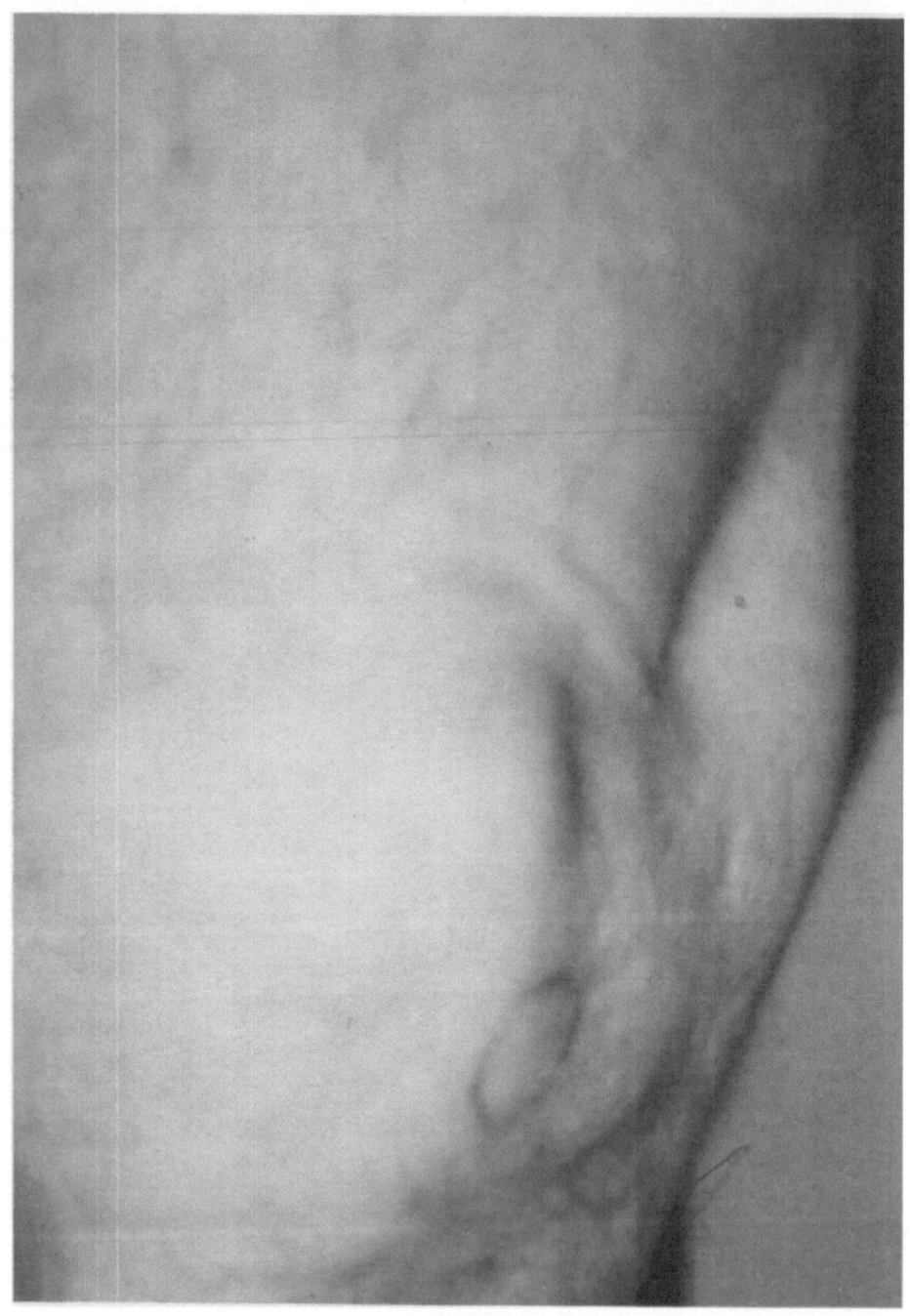

164

7. GONARTHROSIS

Segmental Access: The main supply to the knee is derived from the third, in part also the fourth and fifth lumbar nerve. Impulses travel mainly via the sciatic nerve, sometimes the obturator nerve, and are being brought to the knee by the common peroneal or tibial nerve.

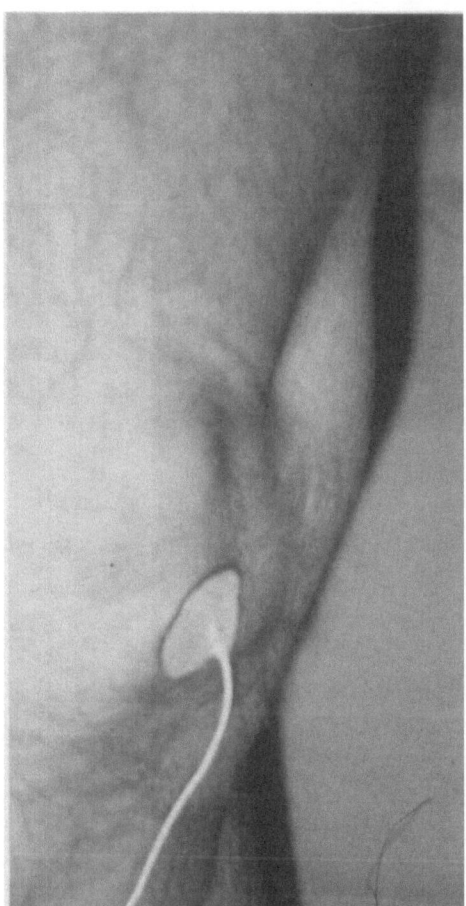

Anatomy: The differentiation which nerve actually will carry the noxious impulses is rather difficult. Exact description of pain location by the patient may help to differentiate (see fig. 16 and 17) but is very seldom provided.

Positions and Sizes of Electrodes: A 7,3 cm^2 anode has to be placed over the medio-cranial aspect of the patella which in most instances represents the most painful site. If the maximum pain is felt at another point place anode there. The large (96,8 cm^2) cathode is to be placed at the opposite aspect of the knee and fixation by tape will be required also if karaya is being used because of the curved surface of the extremity there. – See fig. 93 and 94. Use impulses having higher DC component.

Indications: In all the various types of causation of pain in the knee this type of therapy should be tried. If a specific causation of the pain is known such as rheumatoid arthritis, high uric acid level, gonarthrosis gonorrhoica specific measures must accompany any therapy and also this type of electric therapy for best results.

Fig. 94: Anode in place for gonarthrotic pain of right knee.

Fig. 93: Marking anodal placing for treatment of gonarthrotic pain.

165

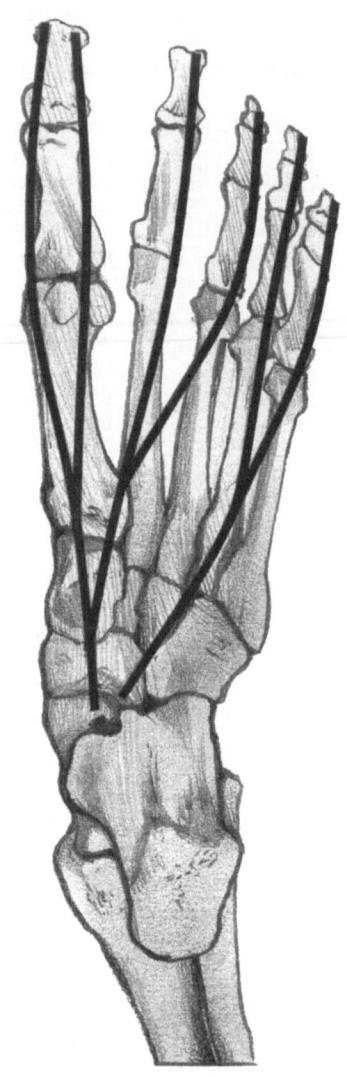

Fig. 95: Anatomic sketch presenting schematically the interdigital nerves of left foot, viewed over skeleton of foot from plantar, medial bundle of nerves innervated from L-5 segment, lateral from S-1.

166

8. MORTON's NEURALGIA

Segmental Access: In this condition one suspects a neuroma of an interdigital nerve (either the second or third most frequently) but very often none will be found. Fibers of the segment L-5 or S-1 are responsible for pain conduction.

Anatomy: It is very difficult to find a very small neuroma or to substantiate a small degree of splayfoot. But it is hard to understand how modern non-invasive techniques like computed tomography or NMR could miss such states. Since several of the patients referred to our pain clinic had operative interventions in vain our opinion is that diagnostic differentiation is rather difficult.

Positions and Sizes of Electrodes: Place the anode ($7,3$ cm^2) over the area of maximum pain at sole of foot. Fixate using karaya. Place the cathode ($96,8$ cm^2) over dorsum of foot with its longitudinal axis in the direction of the interdigital nerve affected. Sometimes the cathode may be only half the given size, then a narrower strip is better than a square shape. See fig. 96 and 97.

Indications: This painful condition needs to be treated extensively. At first an optimal diagnostic work up should try to find reasons for it. The patient (and the surgeon) suffer from any operative procedure if it had been carried out in vain. The difficulties are even increased if one considers that psychic factors sometimes are difficult to grasp. Orthopedic malpositions of the arch of the foot are difficult to identify just as well. A special bandage or arch support for splayfoot in conjunction with electrotherapy may be an optimal combination for most patients and sufficient results are to be expected. An evaluation of the result obtained should be done only after an extensive strain on the foot. A hike of about 4 hours duration without stopping will be sufficient but has to be done supervised: this means that an accompanying person should watch that no breaks are made and that in case of failure before termination of the whole distance the patient may be transported back to the hospital or other institution or his home. No harm should be done in this testing; but objectivation of the endurance should be ascertained.

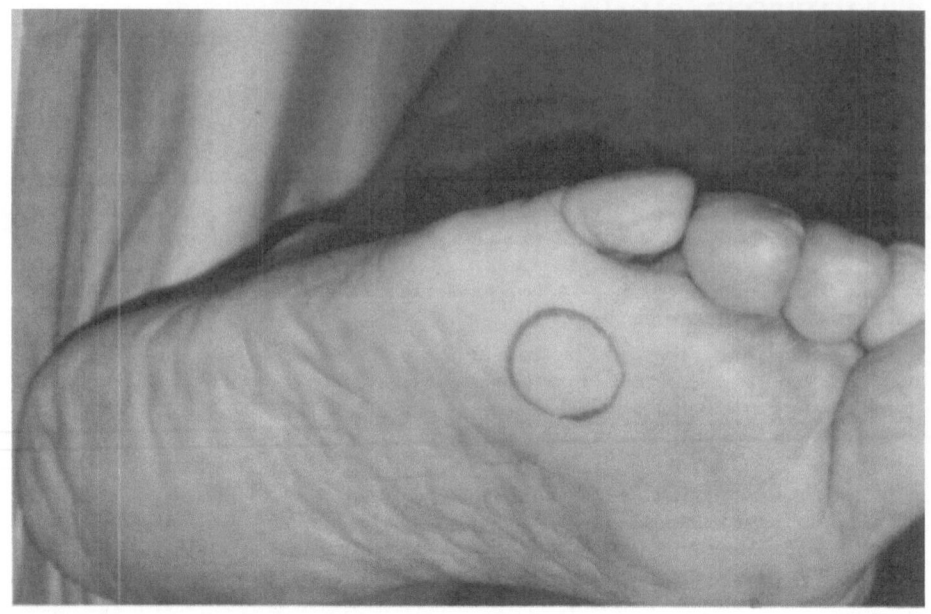

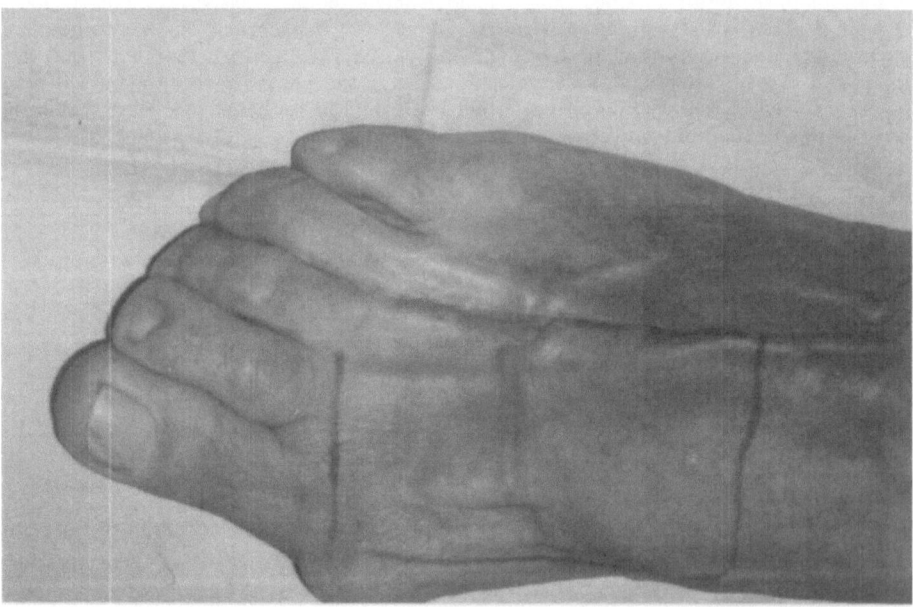

Fig. 96: Marks for anode (upper photo) and cathode (lower) for local therapy of Morton's neuralgia using brief DC-impulses.

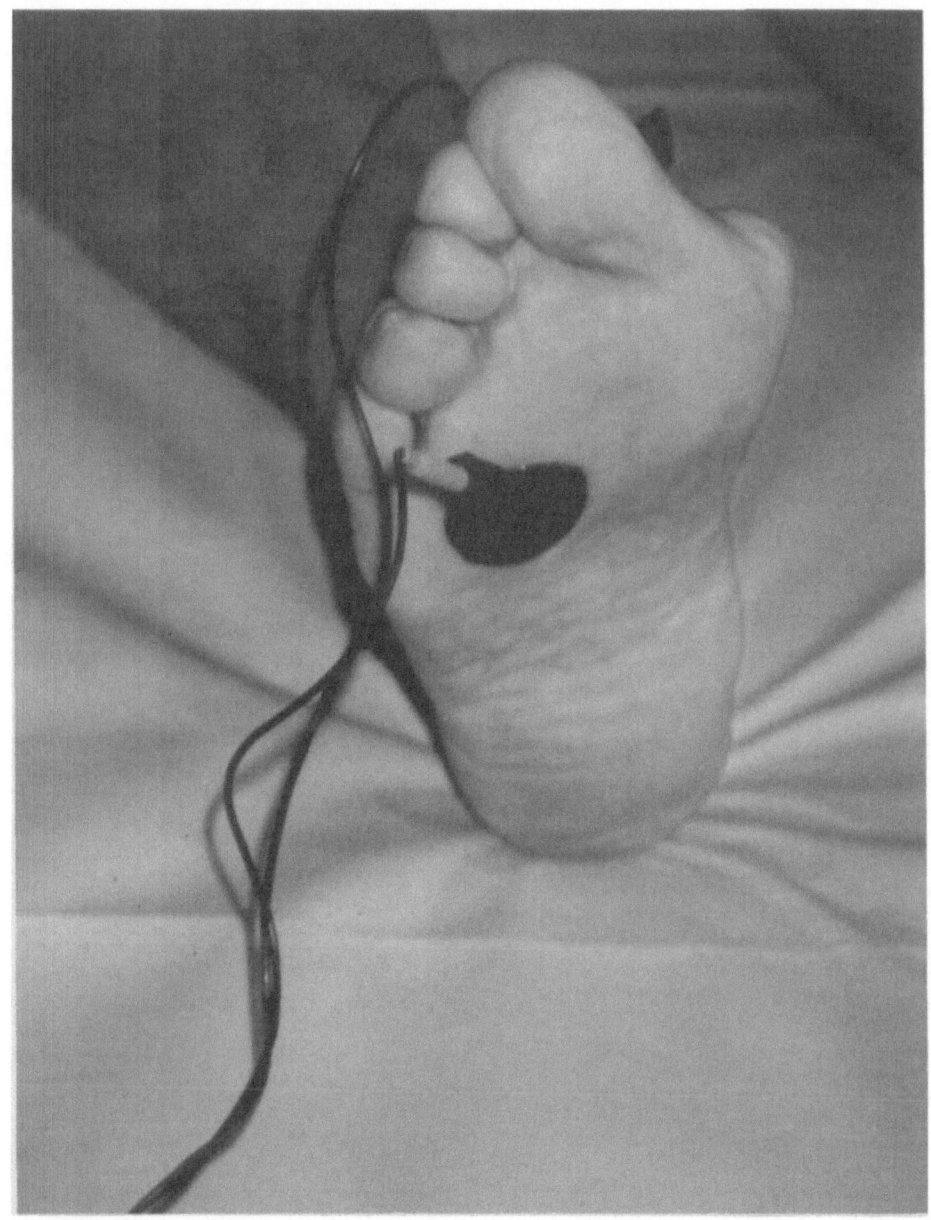

Fig. 97:   Electrodes placed for local therapy of Morton's neuralgia.

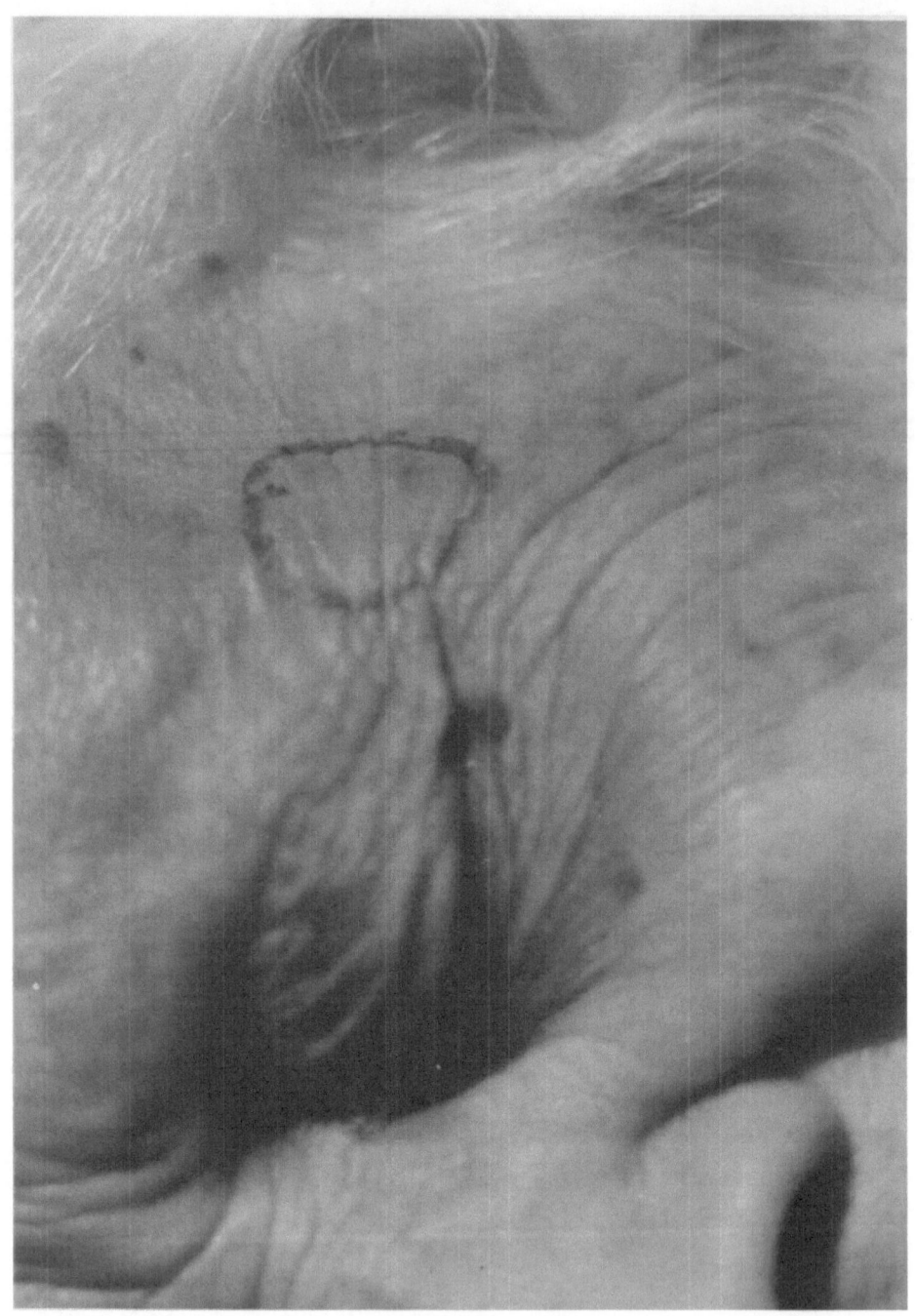

9. ORBICULARIS OCULI MUSCLE; SUPERIOR LEVATOR PALPEBRAE MUSCLE

Segmental Access: Motor innervation of the orbicularis oculi muscle comes from the facial nerve while the levator palpebrae superior muscle derives its motor supply from the oculomotor nerve.

Anatomy: No special remaks seem necessary. Both muscles should be reached and stimulated directly..

Positions and Sizes of Electrodes:A very small (1,3 cm²) cathode (!) has to be placed directly over the muscles of the upper lid. Fixation with karaya is advised. The 7,3 cm² anode (!) has to be placed just laterally of the lateral canthus of the same eye. It should be fixed just over the zygomatic arch as shown in fig. 98. Please note that it is essential to use inversed polarity (compared to use for pain treatments) and the cathode must be the very small electrode! The optimal frequency would be 100 or a bit more (-120) but 35 also suffices while a loger treatment series (twice the number of 20 minute sessions as with a frequency of 100) is required then.

Indications: Unilateral ptosis is the only indication, but only if there does not exist a myasthenic state. It is to be found relatively frequently in elderly diabetic patients. A treatment series of 1 month of daily 20 minute sessions will reverse the ptosis to normal upper lid function .

Fig. 98:   Skin markings for anode (dot) and cathode (ring) to treat unilateral ptosis (overleaf),

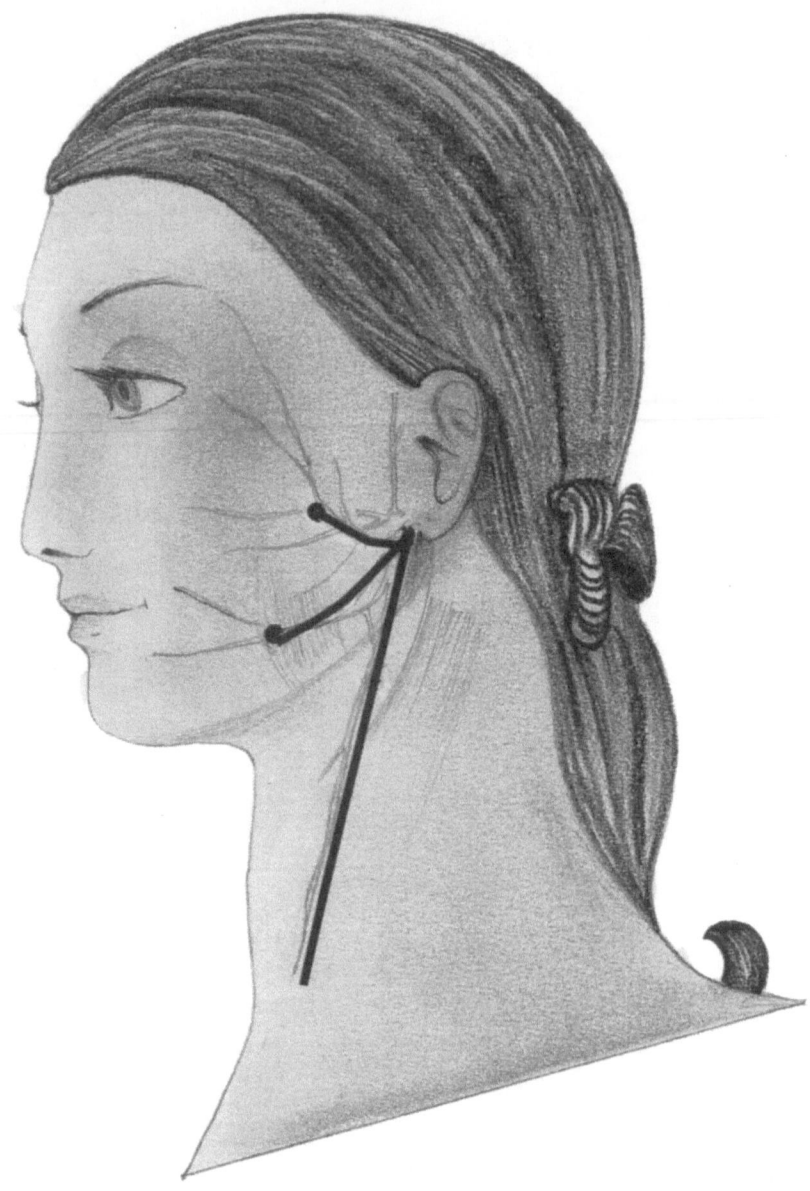

Fig. 99:   Anatomic sketch with schematic presentation of small branches of left facial nerve. Most frequently used positions for (small) anodes indicated by full dots.

10. FACIAL NERVE

Segmental Access: The facial nerve is a largely motor nerve with only a minimal sensory part for the upper-dorsal part of the external acoustic meatus. It has a very short intracerebral part, but a very long portion from its exit from the brain through the petrous pyramid and also a long extracranial stretch. During this course, mainly the transosseous, there exist many possibilities for injuries to the nerve.

Anatomy: Extending the remarks mentioned in the section of "segmental access" the tortuous intra-osseus pathway of the facial nerve makes it susceptible to injuries encountered mainly in the course of fractures of the base of the skull. Also, infectious processes of the inner ear as well as tumors of other cranial nerves in the vicinity of the VIIth nerve or the bone or of extracranial soft parts may affect this nerve and cause damage leading to motor weakness or paralysis of a part or the total motor end apparatus of this nerve. On the other hand there are – mostly of unknown origin – states where the nerve is irritated with resulting tic of the innervated muscles. Not very infrequently these muscular twitchings occur in patients suffering from disseminated encephalomyelitis and developing a demyelinating focus in a part or most of the facial nucleus within the brain stem. The peripheral branches of the facial nerve, which are essential for placing electrodes for electroblock or electrostimulation, are schematically shown in fig. 99.

Positions and Sizes of Electrodes: Whichever kind of therapeuthic use (see under INDICATIONS) may be planned, a very small (1,7 cm^2) and a middle sized (7,3 cm^2) electrode are required. The small electrode always has to be placed over the branch of the facial nerve involved or over the stem if the total nerve is suffering. The larger of the two electrodes has to be placed on the opposite (contralateral) side of the face, usually just under the zygomatic arch (see fig. 100). The polarity of the electrodes depends on the purpose of application and the goal one wants to achieve.

Indications: For treament of a facial paresis (also Bell's paresis) the smaller electrode is to be the cathode and the optimal frequency of impulses would be 1oo/sec. In treating tic the smaller electrode will have to be the anode and the frequency used will have to be 35. For both purposes a treatment session lasts 20 minutes and has to be planned daily. Tic will disappear after about 3-4 weeks, but will recur in about 3 or 6 months. Paresis will diminish after 2-4 weeks and practically disappear (revert to normal function) after between 6 and 12 weeks if daily therapy sessions are consequently done. Function of the facial nerve will recur as long as the nerve has not been totally destroyed. The return of function then will last forever. This is true even for longer lasting paretic nerves such as is shown for a patient (with his consent) in fig. 1o1 and 1o2.

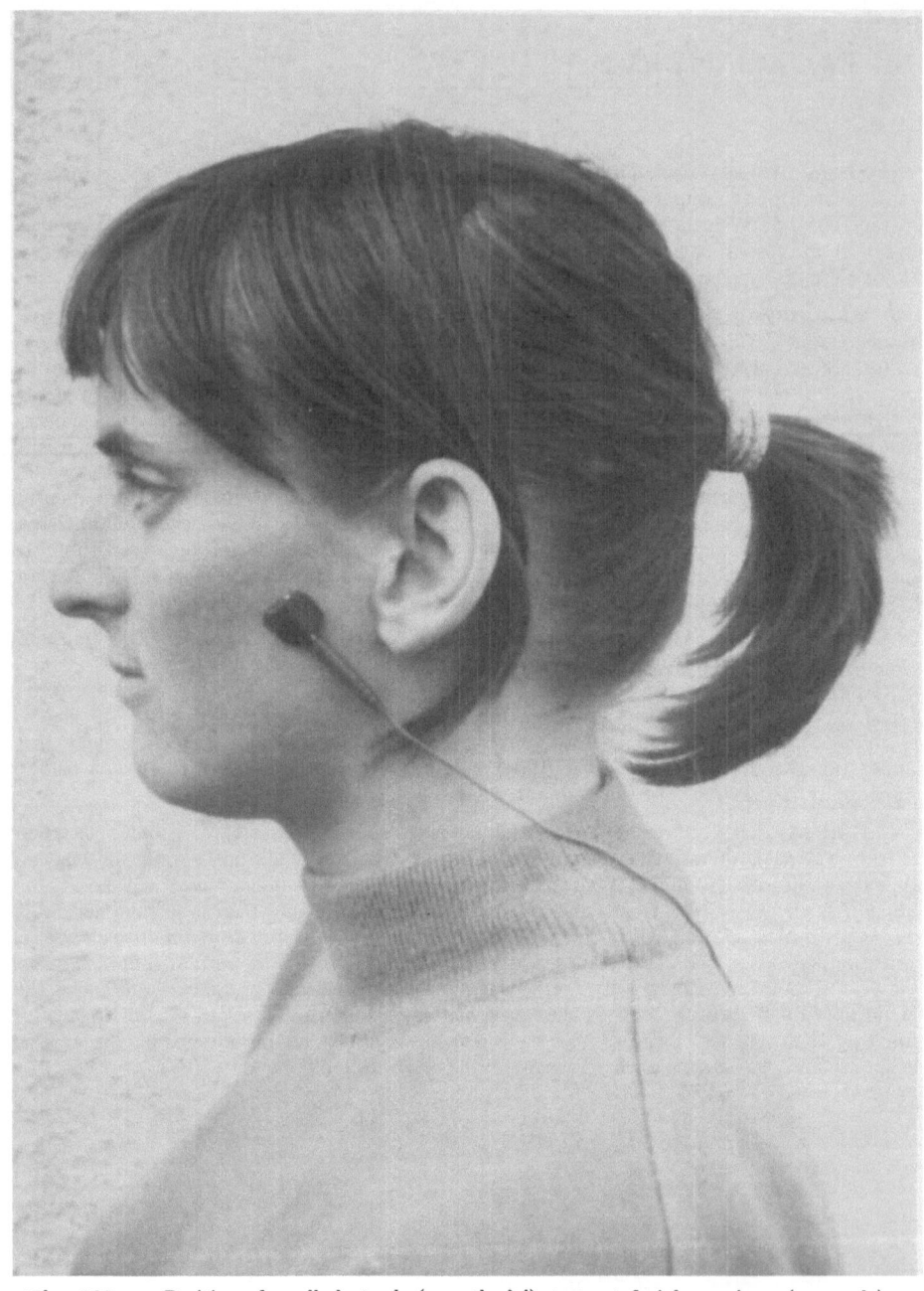

Fig. 100: Position of small electrode (as cathode!) to treat facial paresis or (as anode) in treatment of facial tic of those branches peripheral of small electrode.

174

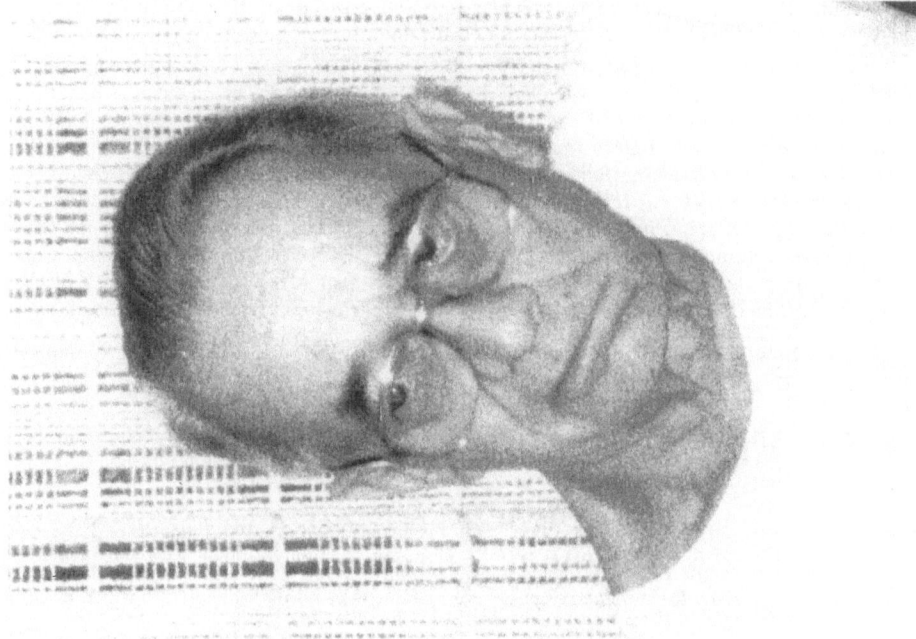

175

11. SUPERIOR LARYNGEAL NERVE

Segmental Access: This nerve is a branch of the vagus nerve. The importance of this nerve for the causation of bronchial obstruction had been shown experimenthally [240]. Also WIDDICOMBE [l.c. 65] thought that for the origin of the reflexes motor and sensory influx is essential. Interruption of the sensory influx would suffice to terminate bronchial obstruction.

Anatomy: The afferent portion of the reflexes mentioned carries impulses of the periphery to the central nucleus of the vagus nerve in the medulla. From there, efferent fibers run along the vagus nerve to the bronchial tree where they liberate the neurotransmitter acetylcholine . This transmitter then causes the smooth muscles to contract and bronchial obstruction ensues. Therefore, attempts to interrupt these fibers surgically had been undertaken. But a double blind study under the conditions to surgically interrupt fibers in the superior laryngeal nerve are not truly possible; the results of the surgical measures were termed equivocal. However, double blind studies with interruption of these fibers electrically are possible. Such a study is being planned after a pilot study on 10 cases (FASSL and JENKNER; [65]) had shown a definte effect on pulmonary function in more than 66% of selected cases. A similar study is already under way (PLATTE[1]). The position of the right superior laryngeal nerve is shown in fig. 103.

Positions and Sizes of Electrodes: An anode of 1,7 cm^2 is placed over the exit of one (or both) superior laryngeal nerves (fig. 1o4). A 96,8 cm^2 cathode is placed over the midline in the neck just below the hair line and down over the uppermost thoracic vertebral spines. This position is correct for either unilateral or bilateral anodal placement.

Indications: The pilot study mentioned [65] selected patients who had been supervised and their lung function controlled without showing too large deviations in either direction. Monthly infusions had been required in all these patients and their medication stayed at a certain level. Before the start and after one and two months following a one month treatment interval the following items were controlled (and the observed changes given in brackets): FVC (+ 16% improvement by therapy); FEV (in % of FVC; after one month therapy -3.8%, i.e. improvement). Subjective impression of patients were such that up to 5o% reduction of medication was observed and even desired, no more infusions were required during the month of therapy and the following month. This applies to 7 of the 10 patients. The other three had initially shown such grave disturbances that no changes

---

[1]Dr. K. PLATTE, head of the general medical department of the municipal hospital Norderney, GFR.

Fig. 101: Privately made photograph of patient before beginning of electric treatment of his facial paresis using brief DC-impulses. Written permission of reproduction obtained. Electrodianostic study showed SPA to have value of 0.

Fig. 102: Privately taken photograph of same patient (as fig. 101) after 3 months of daily therapy at home. Written permission of reproduction obtained. The control of the electric report on sum-potantial-amplitudes was done rather late (on May 21,1990). It showed a reduction compared to the normal side of 72% which in the light of the pre-treatment result of 0 is most remarkable.

took place and, actually, had not been expected. The impression prevails that at least those patients which do not respond well to routine therapy should be given a trial period of this electroblock. More exact indications and data will be expected from the double blind studies under way and planned.

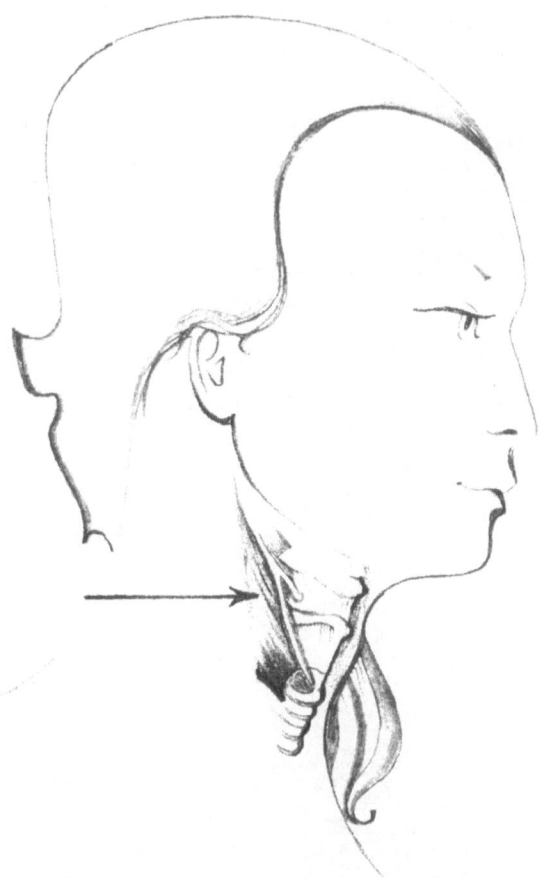

Fig. 103: Anatomic situation of superior laryngeal nerve to find anode in treatment of bronchial obstruction. From FASSL and JENKNER [65].

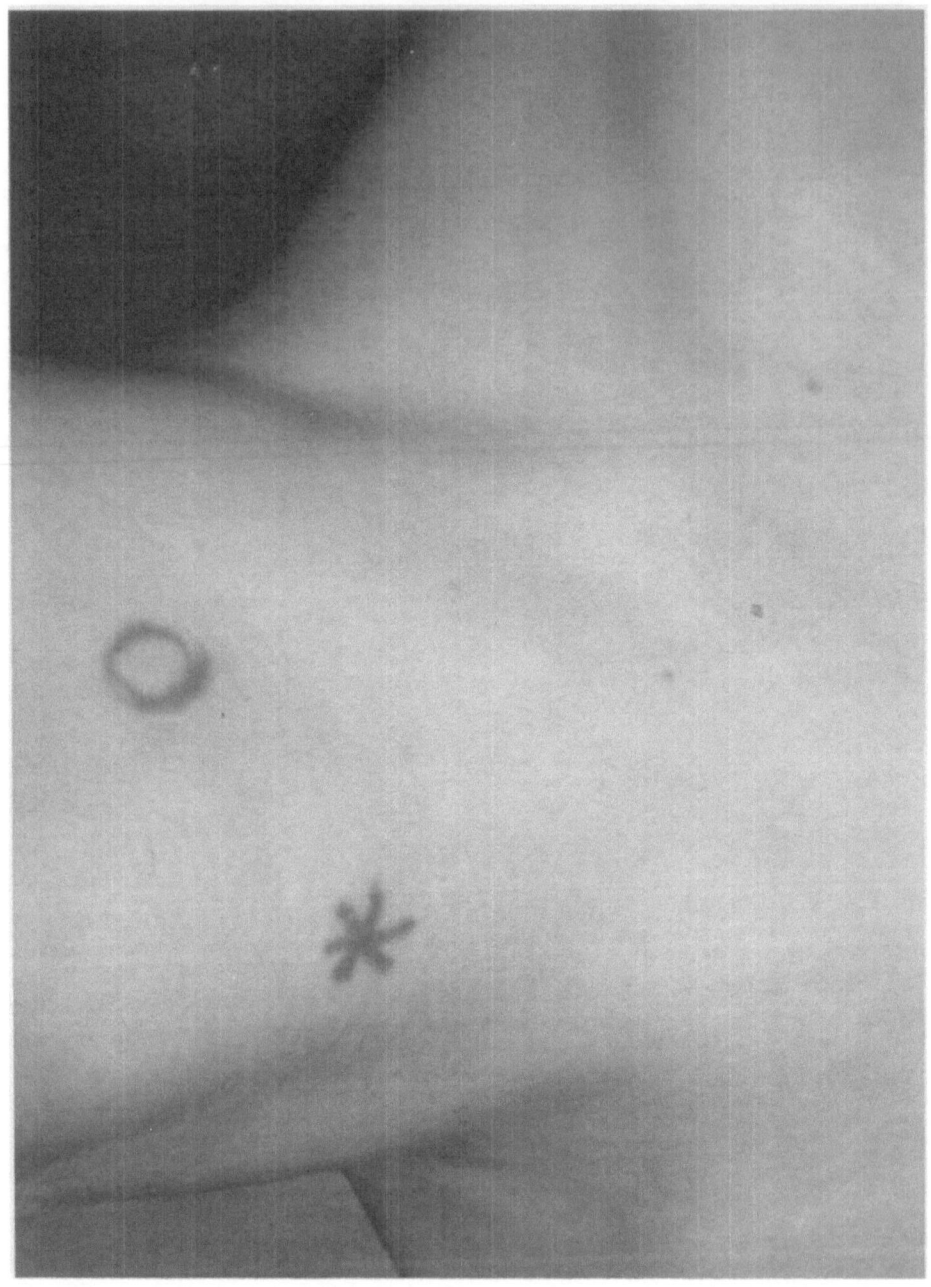

Fig. 104: Mark for anode positions to block electrically superior laryngeal nerve (left). Asteric indicates position of Adam's apple.

12. INSUFFICIENCY OF THE RECTAL AND VESICAL SPHINCTER MUSCLES

Segmental Access: Treatment via the supplying segmental nerves is not possible. Segmental supply is derived from the segments S-2 to Co-1, the prevailing afferents travel via S-3 and S-4 (anal nerve) for the rectal sphincter and for the vesical sphincters the sympathetic segments S-3 and S-4 are relevant.

Anatomy: No special remarks required. The insufficient sphincter muscles have to be in a dense electric field. The impulses to be used are identical to those used in treating painful conditions via nerves, but inversion of polarity is definitely required.

Positions and Sizes of Electrodes: In both instances a specially made electrode has to be used. For the rectal sphincter training a non-conductive rod of the diameter of a small rectoscope is used. In a shape of a ring conductive material (e.g. silver) about 5.8 mm wide covers the surface of this rod at level. This ring is electrically connected through the center of the rod with the cathode (fig. 1o5). This ring then is placed exactly over the mucous membrane covering the inner sphincter. This position has to be taught to the patient and the patient has to experience the sensation derived from correct electrode placement during current flow to be able to ensure correct electrode placement once home treatment is performed. For the insufficient vesical muscle in women a similar rod of non-conductive material has to be used. It has to carry a circular disc (silver) of 5 mm diameter at a certain point (fig. 1o6) which also is connected via the center of the rod to the cathode of the apparatus. In this case the rod should have the diameter of a vibrator. Both rods have to have a conical tip for ease of introduction with the help of some vaseline (electrically non-conductive). The metal disc should be at level with the surface of the rod and has to be placed exactly over the external orifice of the urethra. This position has to be taught to the woman and she has to sense the flow of current to be able to ensure a correct electrode placement in home therapy. For both instances the large rectangular (96,8 cm²) anode should be placed with its longitudinal axis over the linea alba as low as may be practicable on the abdominal wall. In case of much hair a moist pad should be placed between the electrode and the skin. Karaya then may not be used. Best skin contact of the cathode and anode is achieved in a reclining poition, the anode being pressed to the skin by a sand bag (or rice bag which is not liable to become mouldy). For vesical insufficiency in the male patient a circular (7,3 cm²) electrode needs to be placed over the root (or base) of the penis for a cathode and fixation by karaya is easily obtained. The anode has to be placed as mentioned for rectal insufficiency. The frequency of impulses should be 100 for optimal results; 35/sec also gives some results if the number of treatment sessions is higher.

Indications: In any case, insufficiency of the sphincter muscles, be it rectal or vesical, should be subjected to this therapy as long as one is able to document even a minimal function of these muscles. This proof is best obtained by electromyography – but this is rather painful and not easily obtained (if at all). Second choice should be to test the pressure which a sphincter may resist. This should be possible at all urologic departments and is easy. Also in cases where insufficiency is labelled "psychic" this therapy needs to be given to the patient: We know of several patients to whom (after due control of sphincter function) we prescribed this electric training successfully. Daily sessions of 20 minutes are required for three months to obtain about 8o% of return of sphincter function. These patients are very grateful, especially if they had been taking

179

great amounts of psychopharmaca in vain. After another 2 or 3 months (and complete reduction of psychopharmaca) we saw full return of sphincter function.

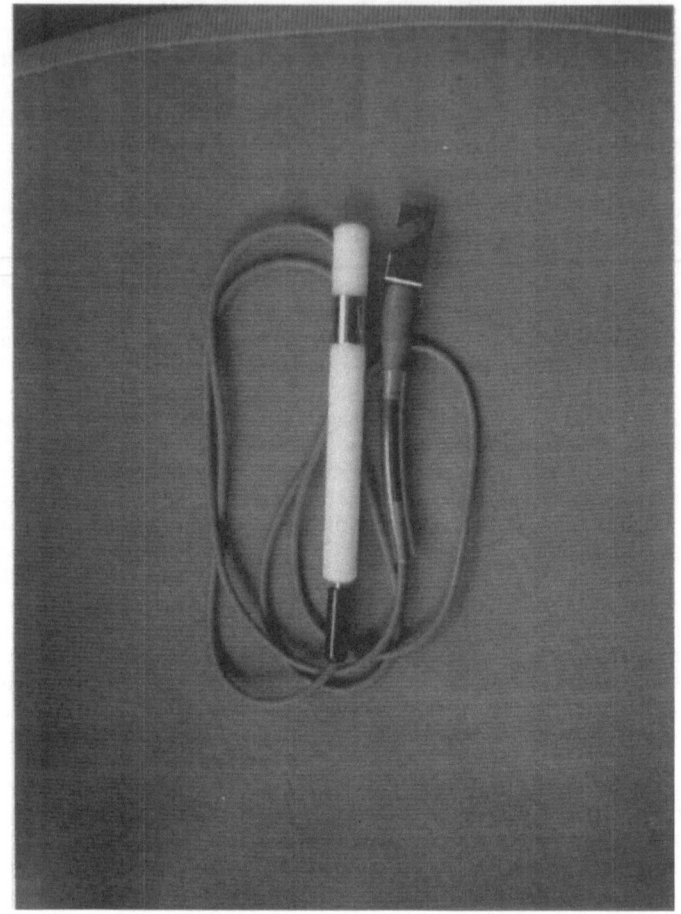

Fig. 105:   Special electrode to treat internal rectal sphincter muscle in rectal insufficiency [124].

Fig. 106:   Special electrode for treatment of vesical incontinency in women by direct electric stimulation of urethral sphincter muscle [124].

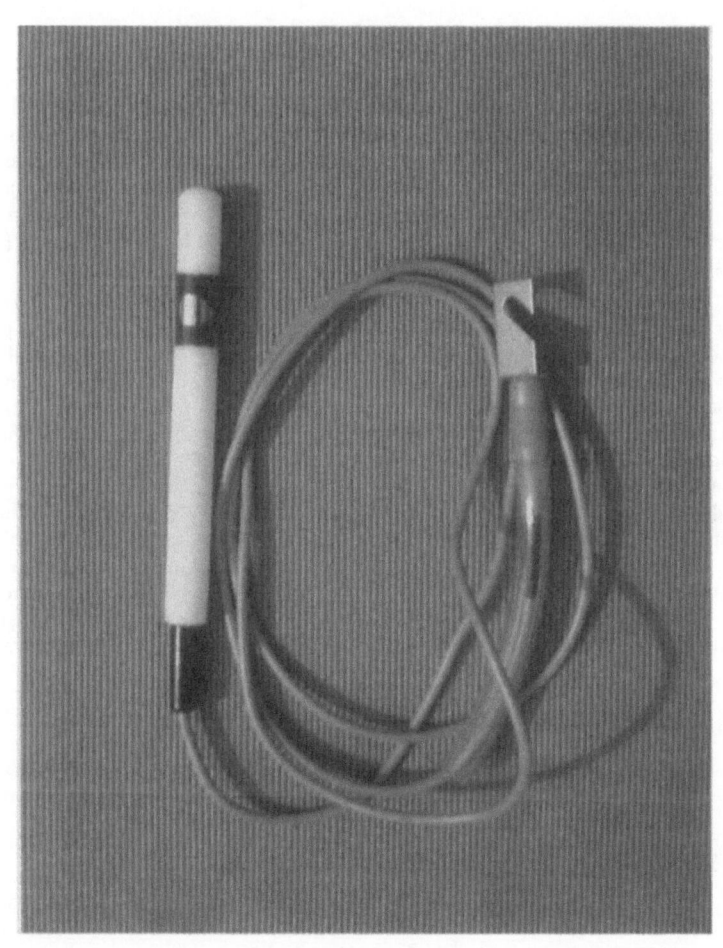

13. MUSCULAR ATROPHY

· In general, muscular atrophy due to a partial damaging of the respective nerve supplying this muscle or group of muscles is to be treated electrically at an institution of
physical therapy. A relatively large apparatus is beeing used for this purpose, and the
therapy ist well established (EDEL, [61]; JANTSCH and SCHUHFRIED, [109]) and effective.
Optimally, this therapy needs to be done daily. If for some reason the patient is not able
to do this or has to be absent from his usual whereabouts for professional reasons (e.g.
business trip), a smaller apparatus would be desirable. The special type of current (or
impulse) needed for this specific use is not usually delivered by any small apparatus listed
in table 2, exept for two instruments. These are ELPHA 2000 by Biometer A/S (DK) and
MonoBLOC by Charters Inc. (USA).

Both these instruments have been tested for this particular purpose and found to
be satisfactory. They provide an important advantage: A part of the routine packaging
provides self-adhesive electrodes needed for this application. The direct muscular stimulation may be combined by stimulation of the respective (supplying) nerve using inverse
polarity (as for pain), of the same form of impulse used for pain therapy. Examples of
both electrode positions are provided in figs. 107–110. Details are given in the legends
of these figures. In these instances, athropy was due to a malpositioning of the patients
on the operating table in such a way that a metal clamp was pressing towards the ulnar
groove and thereby caused damage to the left ulnar nerve. Two weeks after this happened,
ulnar peripheral hypanalgesia was established and muscular atrophy began. This type of
damage is not so very uncommon and has relatively good prognosis if recognized soon and
treated immediately. The type of current to be used should be taken from the description of an apparatus. As stated initially, the routine therapy will stay with institutes of
physical medicine, but for special purposes a small apparatus does have its advantages.

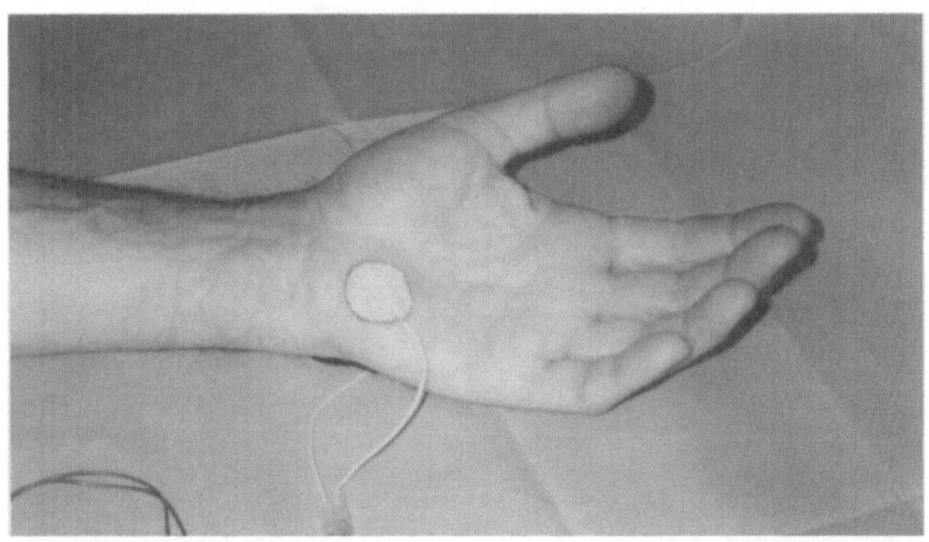

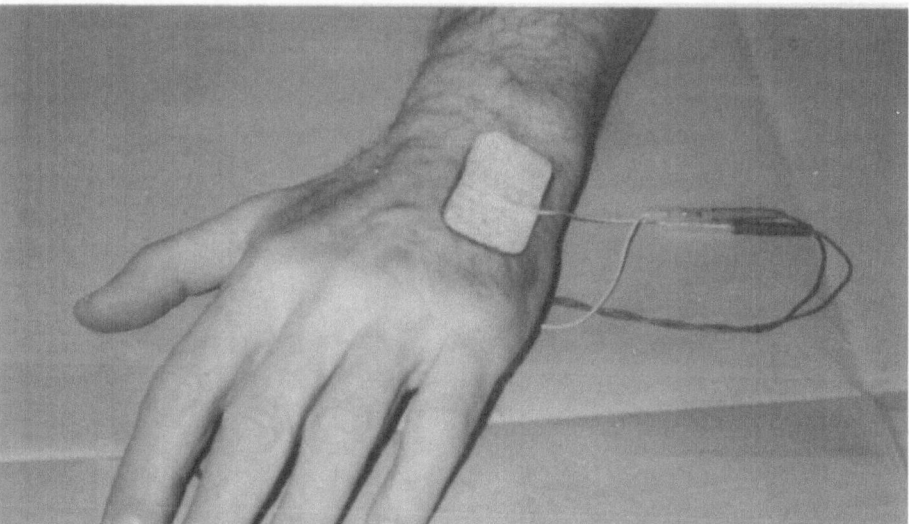

Fig. 107:   Small electrode placed as anode over atrophic muscles of antithenar to treat muscular atrophy.

Fig. 108:   Cathode placed opposite to antithenar for treatment of muscular atrophy of antithenar following damage to ulnar nerve.

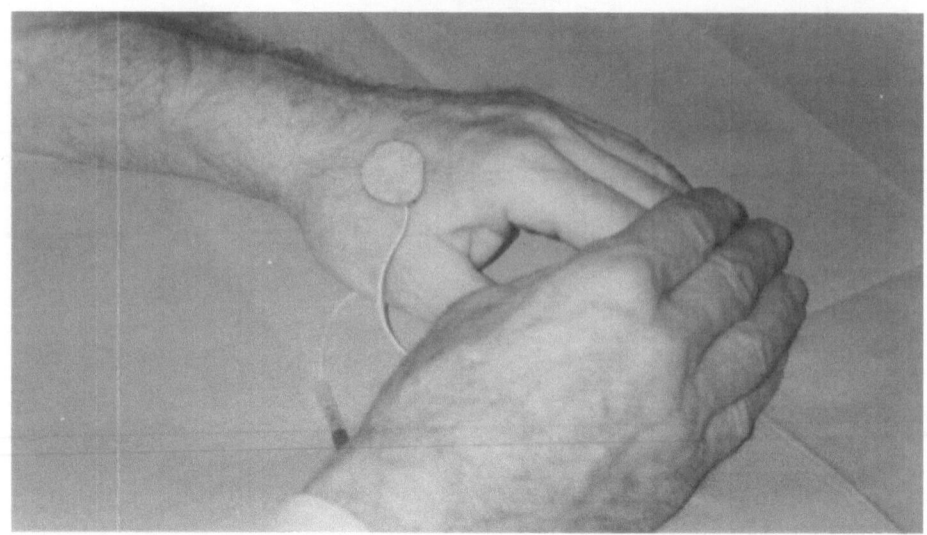

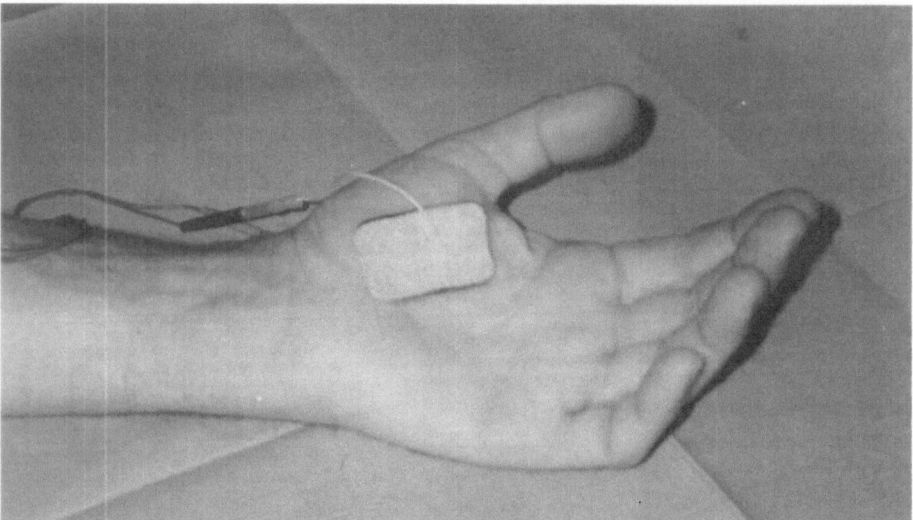

Fig. 109: Anodal placement over atrophic dorsal interossei muscles (following ulnar nerve damage).

Fig. 110: Placement of cathode opposite of anode for treatment of dorsal interossei muscles for atrophy.

ADDITIONAL MEASURES FOR PAIN CONTROL[1]

In using electric measures for pain control – the main feature of this presentation – the reader is reminded that selection of monophasic impulses (pg. 21, 24), i.e. proper equipment, is a prerequisite for optimal effects of therapy. Select the equipment from data presented in table 2 (pg. 26 - 29). Additional measures supporting electrotherapy in various areas or various conditions and various types of pain should be recalled and taken into consideration when establishing a plan for therapy instead of adhering monomanically to electrotherapy only. These measures comprise (a selection only):

(a) MEDICATION:

(1) If electrotherapy ist not succesful enough, pre-therapy medication using a dopamine agonist (see pg. 56) should be recalled.

(2) For migraine a specific substance (fenfluramine hydrochloride = Ponderax, ponderamine) 20 g once daily before bed time (see pg. 70) should be given and attention to the kind of contraceptive (in females of the reproductive age; see pg. 73) is mandatory for optimal results.

(3) Tinnitus and vertigo require simultaneously with electrotherapy an i.v. infusion with metabolically active substances. Directions are to be found on pg. 73.

(4) Specific and documented inflammatory conditions causing the painful state should be treated accordingly (see pg. 162)

(5) Sufficient medication should be prescribed very specifically and **never** p.r.n.! Heavy analgesics should not and need not ever be given.

(b) SUPPORTS: For lowback pain caused by certain conditions, such as spondylolisthesis or postoperatively after disc surgery (for 3 months only to help gradual built-up of activities of daily living) a special back support in the form of a lumbo-sacral corset (see fig. 111) should be considered. It must be custom-made to the measurements of the patient. For a short time, a retention belt (such as e.g. wear'n form) may bring some relief. See also pg. 196. For neck conditions, a soft collar **never** should be used: it would only contribute to loss of strenght and activity of neck musculature; instead, exercises (see appendix, D_2) should be done. To ease pain during sleep or in a resting position, a special pillow (see fig. 112) with a groove for the occiput and a stiff part as a neck support should be used. For both neck and low back conditions, an optimal ...

(c) SLEEPING POSITION should be selected. It should be shown to the patient and corresponds to the position which was initiated by Dick Read for pregnant women. Also exercises prove very helpful. For a brief presentation of these see appendix D_2 and E_2.

(d) MANIPULATIVE TECHNIQUES: These are an essential help for many patients presenting special conditions (such as e.g. whip lash injuries) in the neck or low back. The techniques are briefly described for the most commonly occuring neck conditions (see pg. 94 and 95) and low back pain due to sacro-iliac joint pain (pg. 162). A more precise description of these techniques is to be read elsewhere (e.g. for neck problems: JENKNER [121].

(e) PSYCHOLOGICAL HELP: for all patients should never be forgotten, be it in cases of acute or (even more so) chronic pain. These measures must be carried out by a

---

[1]I am very grateful to M. H. Soalt, D. O. , F. A. A. P. M & R. , Chairman of the Rehabilitation Medicine Department, Orange Hospital Center, N.J., for his suggestion to add this chapter.

185

physician, but should include all relatives (family) and friends of a patient. This already should start immediately after an injury, any beginning pain and taking the case history. It ends only after full recuperation of the patient. BUT NOTE: This should neither be the first nor the only additional measure!

(f) RECOMMENDATIONS FOR A TIME SEQUENCE of the various steps:

(1) Establishing a proper relation (rapport) to the patient.

(2) Reach a correct diagnosis fast; if this takes too long a time, try to find a nerve responsible for conduction of noxious impulses and use electroblocking as a first measure – **but only** if you are sure that this will help; it will not help in case of a whip lash injury, for which problem manual adjustment is required (see under "manipulative techniques").

(3) Decide on plan of therapy and follow it consequently by

(4) whatever measure you deem of optimal effect. EXPLAIN TO THE PATIENT for obtaining his full cooperation!

(5) Check-up after 5 days of therapy to reassure or correct your therapeutic approach. THE PATIENT MUST UNDERSTAND!

(6) Establish a scheme of regular check-ups (either personally or by telephone).

(7) After complete relief, have the patient report in person or by phone every month and after a year personally for a final evaluation.

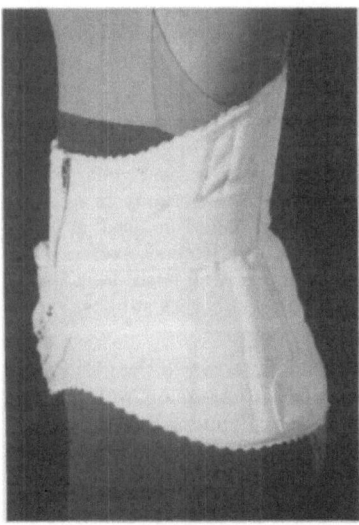

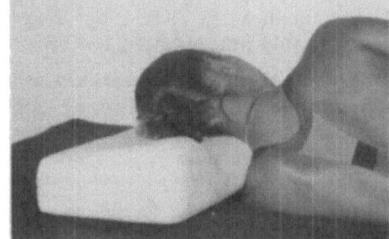

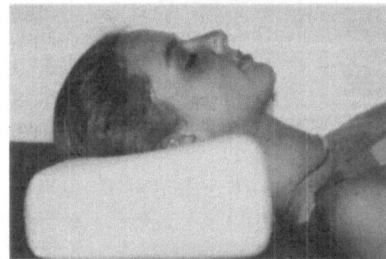

Fig. 111: Patient wearing lumbo-sacral support

Fig. 112: Special pillow for neck conditions.

FINAL REMARKS

Pain I have done with - that I like
(Wilhelm Busch)

The presentation of electric pain therapy given here deviates in some major details from classic electrotherapy. This does not mean that the usual scheme does not act against pain; but we began to investigate an essential prerequisite: the question whether an electric field would act upon nerves at all and if so, which kind of electric impulse was the one allowing to obtain optimal results. After an effect was ascertained optimal current, optimal size and position of electrodes was tested for each condition amenable to this therapy via nerves. If the reader keeps to the specified characteristics of electric impulses and places electrodes carefully where they should be for the respective condition, as described herein, one will obtain the reported results and pain reduction.

I should like to express my thanks and appreciation to all who helped with the concept and manuscript of this monograph: My brother K.W.Jenkner,Ph.D., who revised a part of the manuscript; the artist M.Stelzel for changing available drawings and adding several new ones and Studio Sattler for preparing reproductions. All photographs were made by the author with the explicite consent and cooperation of patients of the "Ambulatorium Süd". My English translation was proof-read by Dr. Ed and Mrs. M. DeMar and Mr T. H. Charters, for which I am very grateful. Last but not least the publishing house be thanked for the tidy production.

APPENDIX

A group of forms used at the pain clinic of "Ambulatorium Süd" for simplification of procedures, to obtain a fast overview on the condition and case history of the patient, for saving time and providing a means of documentation which may also be used to answer questions of insurance companies will be presented. Parts of these which have to be marked by patients were available in several languages. The forms are in sequence to the following list and may be copied by the reader, if so desired, for personal use only. Some of those need enlarging. All forms presented here are suited for computerized documentation and evaluation. Software is not available commercially.

- **A:** General case history
- **B:** Short case history for pain questions, without (B_1) or with (B_2) a possibility of marking site of pathologic lesions
- **C:** Scheme of radicular innervation of skin (from the neurosurgical department of the University of Zürich und the head Prof. Krayenbühl) with written permission of copyright owner J.R.Geigy, Basle, Switzerland.
- **D:** Listing of symptoms for cervical syndrome with space to record test results (D_1) and special exercises (D_2).
- **E:** Listing of symptoms for patients with lumbo-sciatic complaints (E_1) and special exercises (E_2).
- **F:** Pain and weather: notes for 1 month
- **G:** Sequence of examinations for vertigo with space to record results
- **H:** Detailed case history for pain in problem patients (3 pages)
- **I:** Example of scheme of skin innervation to allow physicians to indicate position of electrodes. Should be supplied by producers to physicans or patients.
- **J:** Prevailing electrode positions marked on a segmental skin innervation scheme
- **K:** Size of three standard electrodes which allow to carry out all treatments and obtain optimal results if placed correctly, as observed at pain clinic of "Ambulatorium Süd", Vienna.
- **L:** Self rating depression scale (ZUNG)
- **M:** Head ache questionaire
- **N:** Listing of symptoms of migraine. 15 or more of these establish the correct diagnosis.
- **O:** Pain and weather: desirable blood tests (O_1); Weather factors of importance(O_2).

	Please mark if applicable	yes	no	do not know	since
1.	are you diabetic?	□	□	□	
2.	have you had liver disease or jaundice?	□	□	□	
3.	have you had ulcers in stomach or gut?	□	□	□	
4.	kidney or bladder trouble?	□	□	□	
5.	disease of heart, circulation arteries or veins? high blood pressure?	□	□	□	
6.	diseases of blood, bone marrow or disturbed blood coagulation?	□	□	□	
7.	asthma?	□	□	□	
8.	eye trouble?	□	□	□	
9.	thyroid disease?	□	□	□	
10.	tuberculosis? Other lung disease?	□	□	□	
11.	have you been admitted to a hospital during the last 5 years? Or to a sanatorium?	□	□	□	
12.	have you had an operation during the last 5 years? Which?	□	□	□	
13.	are you allergic? Skin rash? nasal congestion? do you not tolerate some drugs?	□	□	□	
14.	are you pregnant?	□	□	□	how many children
15.	disease of ear, nose, throat?	□	□	□	
16.	diseases of nervous system?	□	□	□	
17.	skeletal system?	□	□	□	
18.	have you had an accident? which?	□	□	□	
19.	how do you sleep?	well □			not well □
20.	do you smoke?	□	□	what? how many?	
21.	do you drink?	□	□	what? how much?	
22.	which drugs do you regulary take?				dosage? tablets daily?

Do you want to give additional information? If so, below:

A plea to my patients!
The treatment you want should be directed to the peculiarities of your disease with its accompanying conditions. In answering questions of this form, you help in making a clinical diagnosis easier and improve my therapeutic proposals.

Thank you for your cooperation.

Your physician

Height 1,__m (__/__); weight ___ kg		
Blood pressure __/__; Pulse ___		
Where do you have pain	Y	N
AREA of HEAD		
SHOULDER/Arm		
CHEST/Breast		
BELLY/Abdomen		
LEGS		

REMARKS OF

PHYSICAN ONLY:

pain index before
after therapy

diagnosis:
referred as

admission dg.

final dg.

therapy plan:

date   signed

signature of patient:

·PAIN QUESTIONAIRE   Nr:

Name	
Address	health insurance?
occupation	hospital

history:

diseases so far ....

....

operations ....

....

irradiations ....

....

pain history:

☐☐☐☐☐

therapy till now: ....

....

effect? ....

proposal now:

....

....

carried out starting: ....

....

....

effect

☐☐☐☐☐

PAIN QUESTIONAIRE   Nr:

Name	
Address	health insurance?
occupation	hospital

history:

diseases so far ....

....

operations ....

....

irradiations ....

....

pain history:

therapy till now: ....

....

effect? ....

proposal now:

....

....

carried out starting: ....

....

....

effect

192

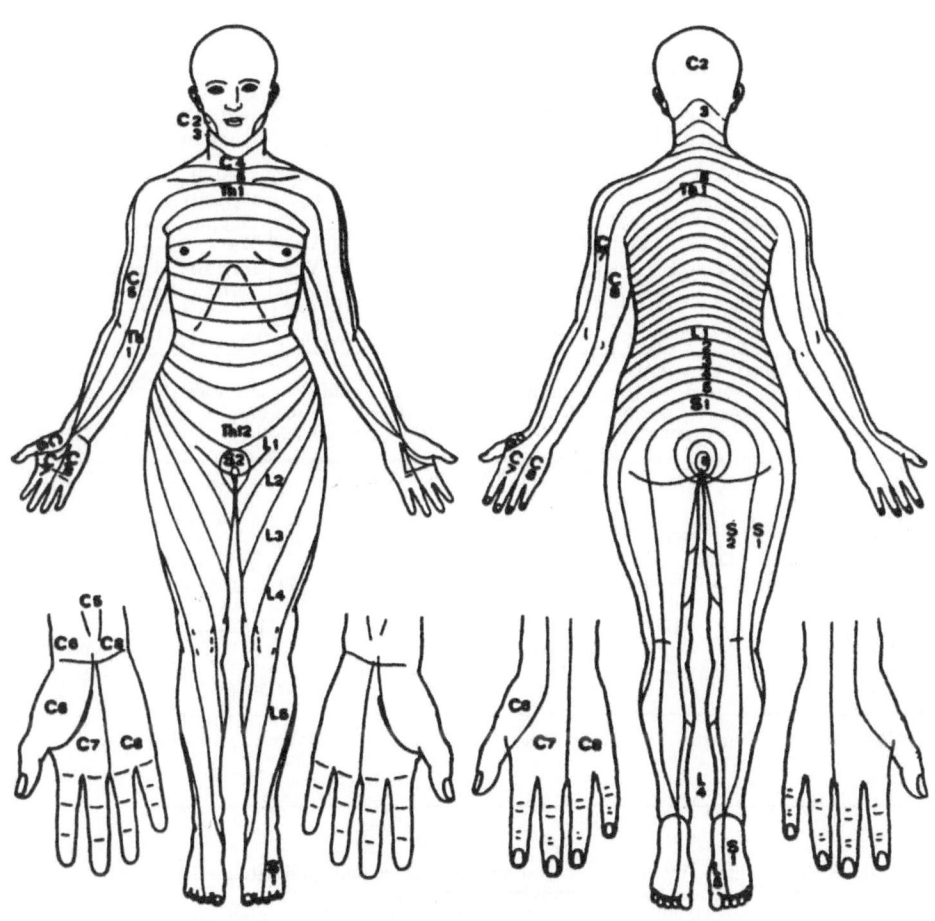

Segmental sensory scheme arrived at and used at "Department of Neurosurgery, University of Zürich". With kind permission of this department and the owner of the copyright, Msg. J. R. Geigy S. A., Basle, Switzerland. Easy to remember are the following levels; C - 2 bordering first branch of trigeminal = hair border on forehead; C - 3 towards 3rd branch of trigeminal = Ear lobe to lower jaw / neck; C - 4 towards Th - 1 = clavicle; C - 6 = thumb; C - 8 = little finger; Th - 4 = nipples; Th - 10 = navel; Th - 12 to lumbar segments = inguinal ligament / fold; L - 5 = great toe; S - 1 = small toe (dorsal, bilateral).

CERVICAL SYNDROME

height   weight
1,_____m   _____kp   Date:_____
___/___   _____lbs

**PATIENTS, please cross out all squares (□) which apply
to your complaints**

Do you have **PAIN?** □no; □some; □moderate; □severe; □most severe.
FOR: □days; □months; □years. HOW OFTEN? □rarely; □occasionally; □often; □always.
ONSET: □suddenly; □gradually; □? MORE SO at □rest, □motion; □stiff neck.
WHERE?

□occiput	C-2	□arm/hand/thumb	C-6	□cold hands	□heart
□collar/neck	3,4	□arm/hand/middle finger	C-7	□sweating	□ears
□shoulder blade	5-7	□arm/hand/little finger	C-8	□numb fingers	□eyes
□shoulder(joint)	C-5	□face □forehead	V/1	□_____	□chest
□upper arm(outwardly)	C-6	□upper jaw	V/2	migraine:□yes □no	
□elbow	C-7	□lower jaw	V/3	vertigo: □yes □no	

MOSTLY?
□mornings; □evenings; □morning stiffness, □joint pains; □sense weather

---

TREATMENT UP TO NOW:

□tablets	□injection	□infiltration	□ultra sound □short waves
□gymnastics	□galvanisation	□cure	□bath □moor
□massage	□accupuncture	□extension	□chircpraxis □other

FOR PHYSICIAN ONLY

Dg.: CS-C_____;   main sympt.:_____;   neurol.deficit:____/____
X-ray: SS   NL   ;KK   C- / ; / ;SL / / ;narrow / ;
□sose   □sart   □ochdr   □DA   □opose
Lab: / ;L ;Ly ;ua ;alc.ph. ;Ca ;P ;Fe ;Cu ;rheuma serol.:_____
□epic.rad.;   □uln.;   periarthr.□h-s   Remarks:
ENT: ECG: ergm.   BP_____
R$_x$: □chpr.;   □exerc.;   □el   □ph.Bl.;
□_____;   □antirh.;   □rh-clinic;   □other(_____)
date: .... check-up: ....

Causes: □sports; □accident; □work □other; □unknown

Result: clin.:   radiol.:

CODE: SS = straight spine; NL = normal lordosis; KK = kink (indicate level);
SL = subluxation; sose = spondylosis; sart = spondylarthrosis;
ochdr = osteochondrosis; DA = dorsal appositions; opose = osteoporosis;
laboratory: space for BSR given by __/__; L = WBC; ua = uric acid;
others self explanatory.

Appointment for: ....
(In case of cancellation please call ....
in advance)

EXERCISES FOR CERVICAL SPINE

I. Head in Neutral Position

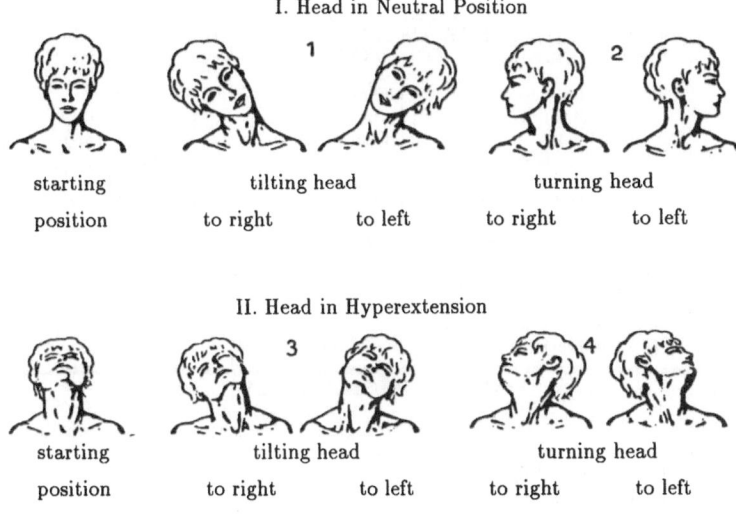

starting	tilting head		turning head	
position	to right	to left	to right	to left

II. Head in Hyperextension

starting	tilting head		turning head	
position	to right	to left	to right	to left

III. Nodding

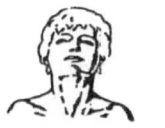 

ATTENTION PLEASE to the following advice:

1. Each exercise should be done 5 times mornings and evenings; as relaxed as possible and not as forceful or extensive as possible.
2. During course of treatment (and not thereafter) please avoid all combined motions of head involving turning and bending at the same time. Avoid during work and sportive activities (such as golfing, tennis, skiing, bowling — these are not to be done during treatment course). For certain jobs work may have to be interrupted for two weeks, such as in case of auto mechanics. When driving do not turn head, use mirror only for rear view. In house hold chores, bedding, kitchen work or care for small children, attend to avoiding combined movements as indicated above!
3. After end of treatment, as well as during same, do not circle your head (as is done frequently during gymnastics) and do not lie or sleep on your stomach (flat, neither during sun bathing). Due regard is advised for full recovery and lasting well being!

195

LOW BACK PAIN

height   weight

1,____m   ____kp   Date:_____

____/____   ____lbs

**PATIENTS, please cross out all squares (□) which apply
to your complaints**

Do you have **PAIN?**   □no;   □some;   □moderate;   □severe;   □most severe.
FOR: □days; □months; □years. HOW OFTEN? □rarely; □occasionally; □often; □always.
ONSET: □suddenly;   □gradually;   □?   MORE SO at   □rest,   □motion;   □cough.
WHERE?

□back	□mid line	□left	irradiates:
□hip	□kidney	□right	□up-ward
□groins	□flank	□outward	□down-ward
□thigh	□gluteal mm.	□inward	unable to:
□leg	□heel	□front	□bend forwrd
□foot/knee	□sole	□back	□stand uprt

MOSTLY?
□mornings;   □evenings;   □morning stiffness,   □joint pains;   □sense weather

---

TREATMENT UP TO NOW:

□tablets	□injection	□infiltration	□ultra sound	□short waves
□gymnastics	□galvanisation	□cure	□bath	□moor
□massage	□accupuncture	□extension	□chiropraxis	□other

FOR PHYSICIAN ONLY

Dg.:_____;   leading symptom_____;   neurol.deficit:____/____
finger-floor-dist.:   ext.rotation □   M.quadr.(L3,4)   □ +   □ – │ □ gyn.
AJ + (rt,lt) – (rt,lt)   standing on heel (L5)   □ +   □ – │ □ urol.
KJ + (rt,lt) – (rt,lt)   standing on toes (S-1)   □ +   □ – │
Naffziger + –; Lasègue + – rt/lt; sensory: □L1; □L2; □L3; □L4; □L5; □ S1; □S □P
painful: □ spine □ SI-joint □ vasc.neur.compt.thigh □ calf:pulse + –
X-ray: SS   NL   ;narrow   /   ;   /   ;wide   /   ;scol.   /   ;
□tors   □o-yt;   listh.   /   ;□o-osis
□ CT   ;□ M–y   ;□ NCT
Lab: BSR   /   ;L   ;Ly   ;ua   ;alc.ph.   ;Ca   ;P   ;Fe   ;Cu   ;rheuma serol.:_____
Remarks: skin temp. ____ rt/lt;
ENT: ECG: ergm.   BP_____
R$_x$: □chpr.;   □exerc.;   □el   □ph.Bl.;   □antirh.;   □other(_____)
date: ....   check-up: .... □corset; □heel-add.

Causes: □sports;   □accident;   □work   □other;   □unknown

Result: clin.:   radiol.:

FORM for LOW BACK PAIN and SCIATICA:
simplified documentation for quick reference of complaints as well as neuro-orthopedic findings.
Space is provided for name, address, social sec.nr., height, weight, symptoms and therapy so-far.
Documentation for initial diagnosis, main symptom, neurologic deficit; S = sensory loss, P =
paresthesias. X-ray report: CODE: SS = straight spine; NL = normal lordosis; CT = computed
tomography; M–y = myelogram; NCT = nerve conduction time; segmental data for wide or
narrow interspace to be entered; scol = scoliosis, giving level of vertex; tors = torsion; o-yt =
osteophytes; Listh = listhesis (segment!) o-osis = osteoporosis; ECG = electrocardiogram; ergm
= ergometry; chp = chiropractic; exerc = exercises (see E_2).

Exercises for lumbar spine

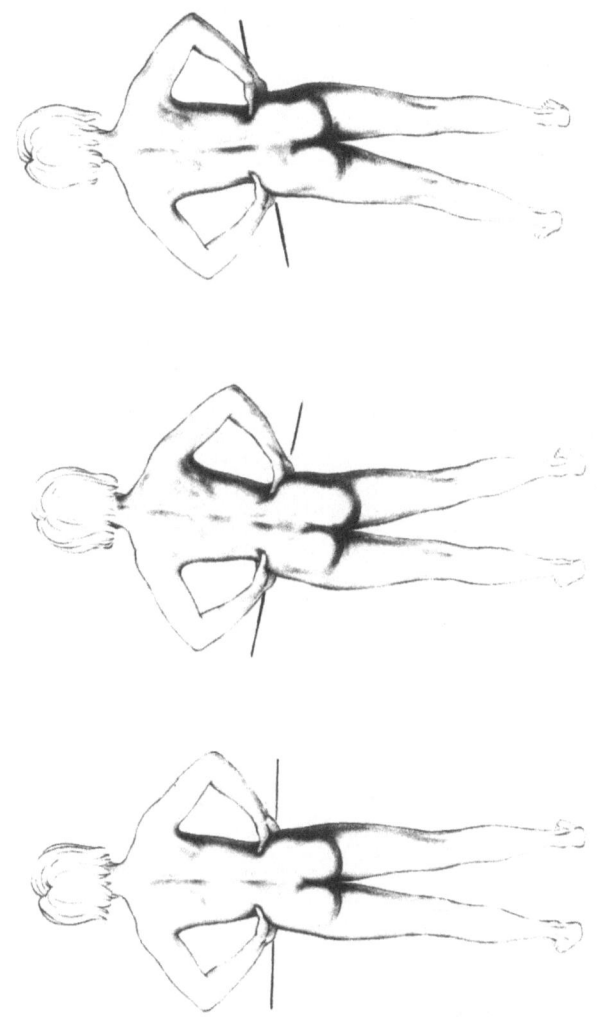

These exercises are carried out standing, feet somewhat apart, hands placed over iliac crests, arms relaxed. Starting with **left** position, where iliac crests are horizontally, pelvis has to be tilted without bending knees simply by changing weight to left leg (**center**) and right leg (**right**). Body and thoracic spine stay in vertical position.

Second exercise consists of tilting pelvis in another way, namely ventro-dorsally: push iliac crest fore- and downward (**far left**), then up and back (**center left**). The latter is helped by leaning towards a wall, keeping feet a foot away from wall.

Lastly (third exercise) change to a sitting position again tilting pelvis ventro-dorsally. Change from ventral down (**center right**) to (**far right**) dorsal down.

Do every exercise five times in the morning and five times before bedtime.

198

pain and weather

name: ....
address: ....
....

month: ....
sex: ....
age: .... (. . / . . / . .)
height: .... weight: ....

	1.	2.	3.	4.	5.	6.	7.	8.	9.	10.	11.	12.	13.	14.	15.
severe pain															
moderate pain															
slight pain															
no pain															
vertigo															
menstruation															
stress															
weather															

	16.	17.	18.	19.	20.	21.	22.	23.	24.	25.	26.	27.	28.	29.	30.	31.
severe pain																
moderate pain																
slight pain																
no pain																
vertigo																
menstruation																
stress																
weather																

$\boxed{G}$

MAIN SYMPTOM : VERTIGO   present since ....

accompanying symptoms: ....

. ...

Referral to rheoencephalography is only possible with all reports mentioned on this form.

Blood pressure   /   mm Hg   vascular cause is

ECG: ☐ possible

☐ impossible

acoustic nerve test: ENT cause

vestibular test: ☐ possible

☐ impossible

X-ray report on cervical spine:

(films to accompany patient!)   space for patient identification

==

l e a v e   b l a n k !

Diagnosis: X-ray control

Therapy: EEG

begin: .... ending: ....   Doppler

Summary: scan

angiograms

REG-appointment: (1) ....   **follow-up:** ....

$\boxed{H_1}$

Name: .... born .... date ....

In the body scheme on next page (H_2) please mark pain areas visibly and designate as (1) the most severe pain; lesser pain as (2) and the even lesser as (3). If pain is being felt deep, mark **D** next to number; code superficial pain as **S** next to number. Mark irradiating pain with lines to areas, where pain is irradiating to. In the small table, enter % data on severity of pain giving a number for each of the different kinds of pain you are sensing:

scale:	0 - 20	20 - 40	40 -60	60 - 80	80 - 100
pain:	little pain	some pain	moderate pain	severe	very severe
1					
2					
3					

Give the overall average of your pain as .... %.

Give the degree of pain you are having while filling in this form: .... %.
USE ABOVE SCALE !

when it was the severest ....
when it was very little ....
the severest toothache you ever had would rate as ....
the severest headache would rate as ....
the most disturbing stomach ache you ever had would rate as ....
the most unpleasant sun burn you had would rate as ....
the most annoying insect bite you had would rate as ....

You had probably seen various physicians for your pain. If so, indicate which you saw:

general practitioner ☐
specialists for medicine ☐ | surgery ☐ | orthopedics ☐
                 dentist ☐ | ears ☐ | neurologist ☐
                 gynecology ☐ | other: ☐

You may have seen other people e.g.:
chiropracter ☐ ; masseur ☐ ; priest/clergyman ☐ ; or: ....

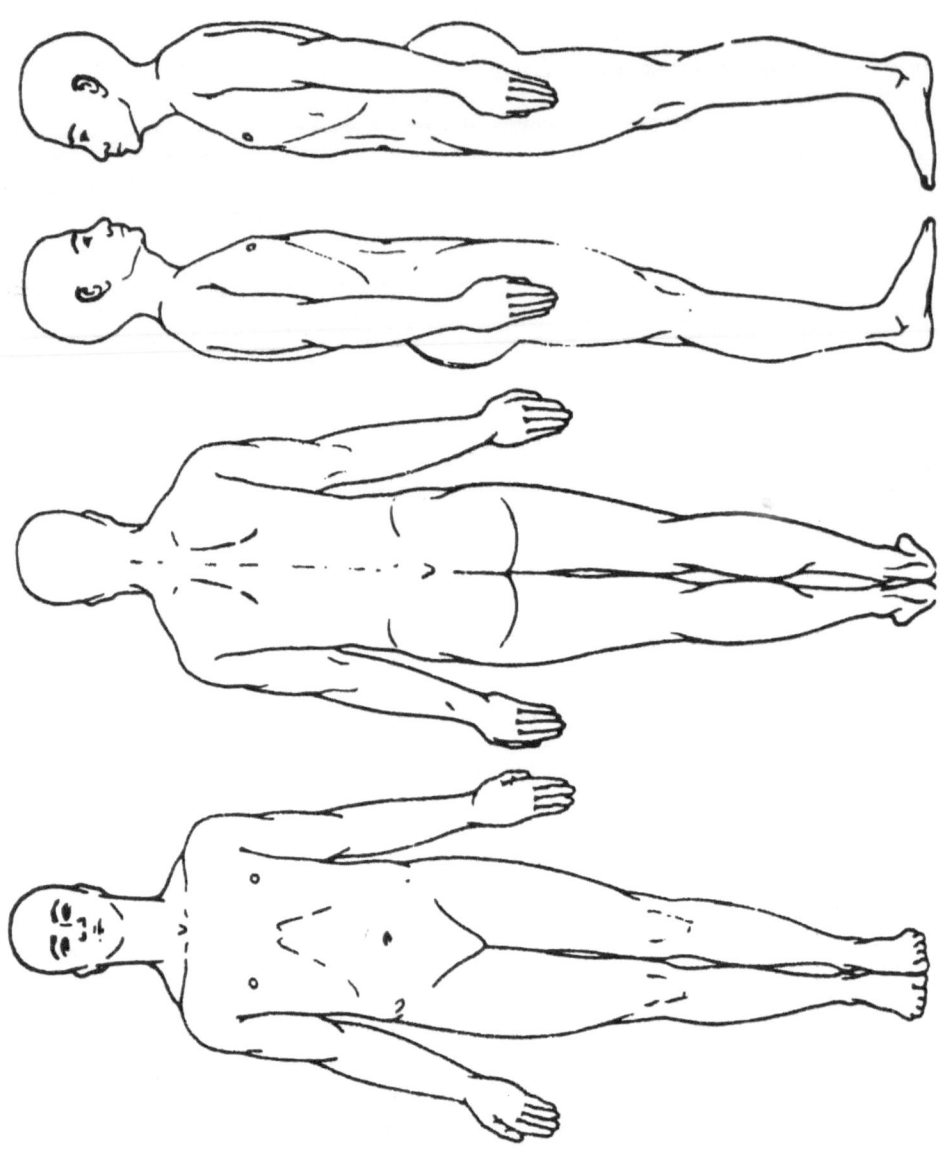

If you compare your life (now, with pain) to your life before you had this pain, you may see that some things have changed like: your **ability to work:** □ no; □ a little; □ much; □ very much

Considering **social life** like: meeting friends, going out etc.	the **desire** for this is: □ identical □ somewhat less □ markedly diminished □ entirely lacking	the **ability** for it is: □ identical □ somewhat less □ markedly diminished □ entirely lacking
your hobby: (sports, hiking, reading)	**desire** □ identical □ somewhat less □ markedly diminished □ entirely lacking	**ability** □ identical □ somewhat less □ markedly diminished □ entirely lacking
your sexual life	**desire** □ unchanged □ somewhat less □ markedly diminished □ entirely lacking	**ability** □ unchanged □ somewhat less □ markedly diminished □ entirely lacking

your sleep habits and ability to sleep:	□ identical
	□ disturbed; insofar as I can not
	  □ fall asleep
	  □ sleep only a short interval
	  □ have to take sleeping pills
		
		□ wake up frequently
		□ do not sleep

Tell, which activities change the sensation of pain to the worse (then indicate by "+" or if the pain is lessened, then indicate by " –".

□ meals	□ motion/walking	□ sitting up
□ drinking coffee/tea	□ rest/lie down	□ coughing/sneeze/bowel motion
□ drinking alcoholic bev.	□ massage	□ sport or gymnastics
□ warmth	□ pressure	□ psychic stress
□ cold	□ bright light	□ physical stress
□ humidity	□ loud noise	□ distraction (TV, babies, etc.)
□ changing weather condition	□ work	□ sexual intercourse

Under which conditions did your pain start (of course only, if you recall):

□ after an accident	□ after a disease (which?)	□ after sleep
□ after an operation	□ I do not see a connection	□ other circumstances

When did your pain start at all? ....
When was your pain very severe? ....
Is the pain always of the same degree □ or sometimes more severe and sometimes less severe? □
What do you do if pain becomes very severe? □ nothing

□ take a drug	□ void	□ walk about
□ lie down	□ take a warm shower	□ take a cold shower
□ other:		

Tell which treatments you already have had: indicate by "+" if it was successful or "0" if not; if it made your pain worse, indicate by "–".

□ drug (which?)	if necessary, please make a listing!		
□ injection	□ i.v.infusion	□ massage	□ galvanisation
□ skin wheal	□ acupuncture	□ short wave	□ iontophoresis
□ infiltration	□ acupressure	□ ultra sound	□ TENS
□ nerve block	□ physical therapy	□ hot packs	□ other (which?)

Had you been operated upon because of your pain: □ once □ more often? □ no.   if yes, do you know which operation was performed? Name: .... date: ....
Without an operation, had you been admitted to a hospital? □ yes □ no

What do you feel so far acted best on your pain: ....

Have you been satisfied by therapy so far? □ yes □ no
The physician's attention to you was ok; □ yes □ no
If you answered no, please explain ....

Did someone say your pain was all in your head? □ yes □ no .
You will have to live with your pain: □ yes □ no .

Patient's signature:

Patient Instructions
Electrode Placement

Patient: .... Date: ....

Therapist: .... Clinic: .... Phone: ....

Special Instructions: ....

Next telephone contact (to nr ....): ....

Next office visit: ....

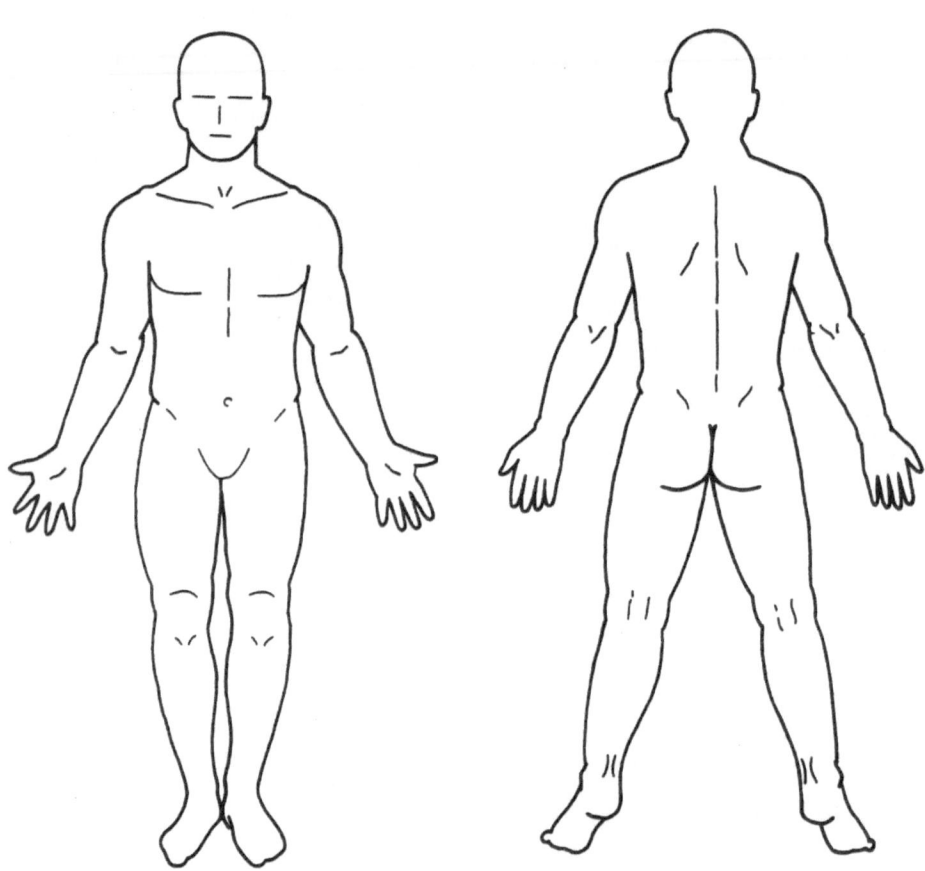

Position of electrodes:
× small electrode (plus for treating pain)
○ large electrode (minus for treating pain)

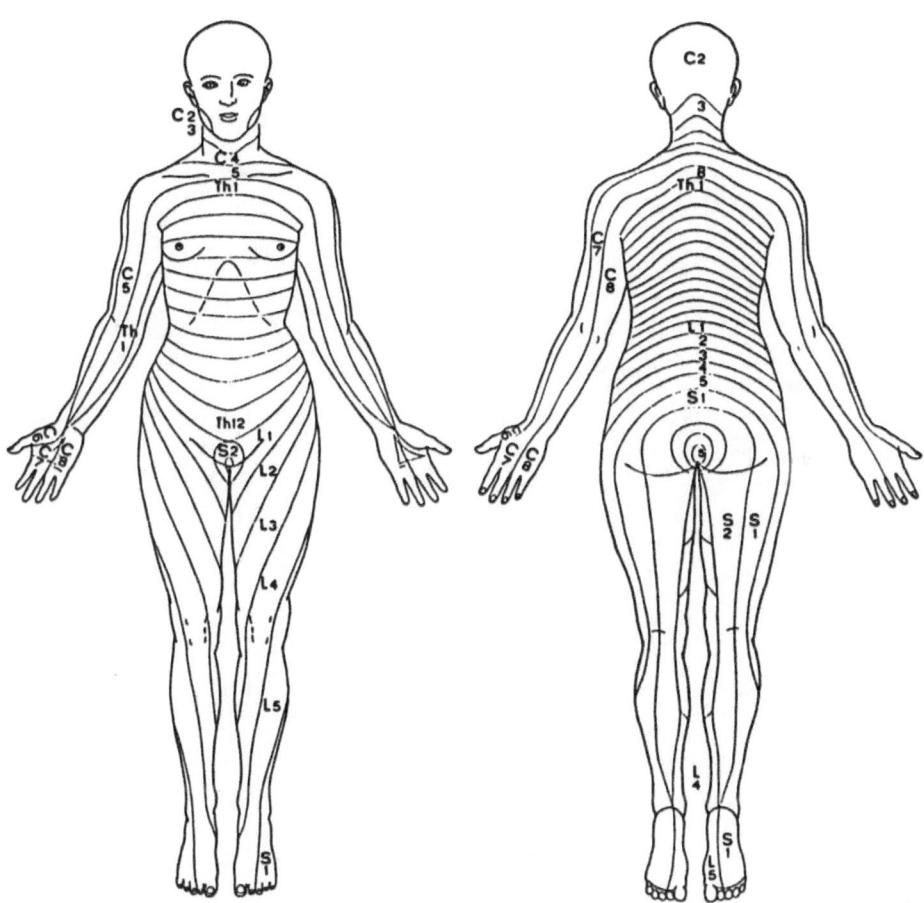

name: .... first name: ....

address: ....

start of therapy on: .... end: ....

rate: .... X daily, ... X weekly

duration: .... min, gradually incr. to .... min.

check up on: ....

report: ....

....

$\boxed{J}$

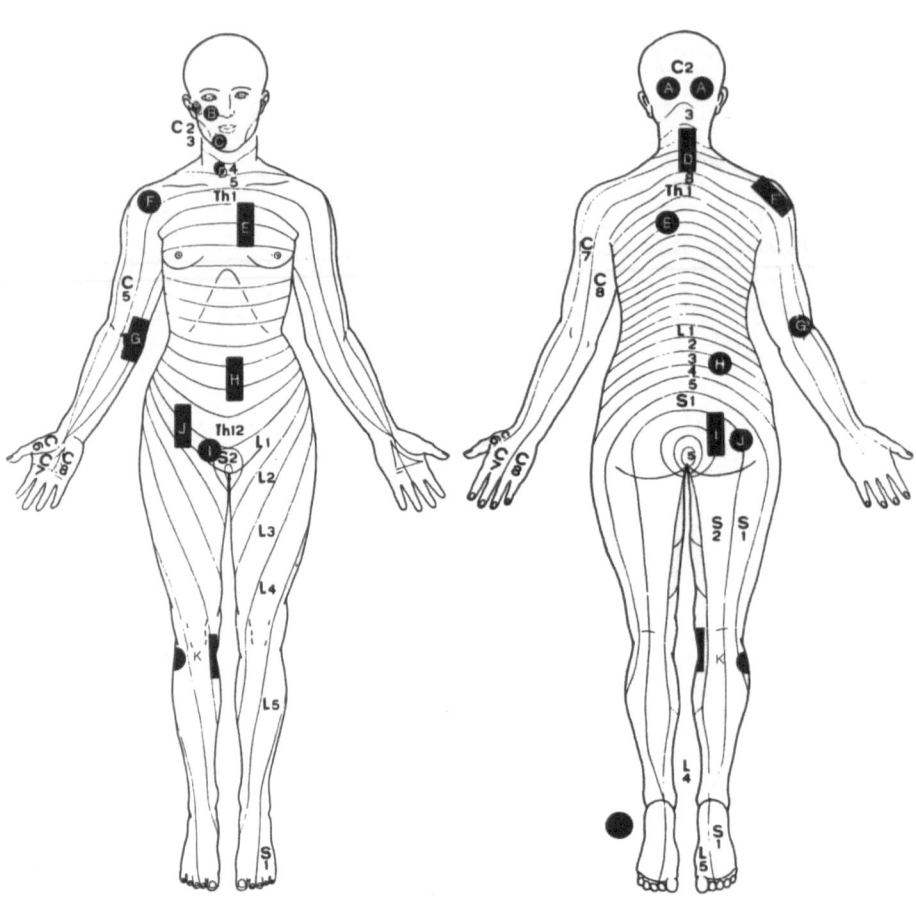

206

Electrode positions for:

A   occipital neuralgia
B   trigeminal, 2nd branch
C   trigeminal, 3rd branch
D   stellate block
E   intercostal block
F   periarthropathy hum. scap.
G   epicondylitic pain
H   lumbar sympathetic
I   obturator nerve
J   sciatic nerve
K   peroneal nerve

Standard electrodes allowing all possible treatments
giving their size and purpose.

Electrodes	size	area	type of application
A	1,4 cm ⊘ 0,56 in ⊘	1,6 cm² 0,25 in²	anode for stellate block, trigeminal neuralgia and some rare neuralgias
B	3,0 cm ⊘ 1,2 in ⊘	7,3 cm² 1,1 in²	anode for most blocks; cathode for trigeminal neuralgia and some rare neuralgias
C	7,6 x 12,7 cm 3 x 5 in	96,8 cm² 15 in²	cathode for all blocks unless **B** is being needed

Use similar sizes of commercially available self-adhering reusable electrodes such as e.g. PALS PLUS (Axelgaard M.Co.Ltd.), UNI-PATCH (Uni-Patch Inc.) or Electrodes for Nerve Stimulation (Medicom A/S, DK-3650 Olstykke, Danmark) and cut down if no real small size is made available.

Attention: Skin oils, lotions and cosmetics on the skin will prevent all self-adhering electrodes from sticking tightly. Therefore, the sites selected for applying electrodes should be free of such oily substances; to clean the skin soap and water or alcohol wipes work well. If oils etc. get on the sticky surface of self-adhering electrodes, remove these with an alcohol wipe. Keeping skin and electrodes free of oil, lotions and cosmetics will greatly extend the life of self-adhering electrodes.

Small Round
0.56" diameter

Medium Round
1.2" diameter

Large Rectangle
3" x 5"

Self-rating Depression scale
W.W.K.Zung

This self-rating scale serves for quick evaluation of depressive states. The examined patient should understand what to do: A cross should be made in the respective column for each statement.

Of the 20 statements, 10 are formulated to give typical symptoms occuring in depressive patients. The 10 others are selected to present typical life situations which are usually disturbed in depressive patients. For evaluation of answers, place cellophane with imprinted figures exactly over form and write value of answer in column provided on far right. Summing up the 20 values give the raw score.

These raw values indicate lack of depression if sum keeps under 40. 41 - 47 indicate light depression, 48 - 55 mean moderate to severe depressive state and over 56 a severe depression is to be accepted. The author ZUNG introduced an SDS-Index by dividing the sum of the raw score by 80 and multiplying it by 100. Evaluation by index must have the limits for the various depressive degrees at 25% higher than the raw score.

The validity of this test has been compared by various other methods of testing (e.g. Hamilton Depressive Scale or MMPI).

Name	None or a Little of the Time	Some of the Time	Good Part of the Time	Most or All of the Time	
Age .... Sex .... Date					
1. I FEEL DOWN-HEARTED, BLUE AND SAD					
2. MORNING IS WHEN I FEEL THE BEST					
3. I HAVE CRYING SPELLS OR FEEL LIKE IT					
4. I HAVE TROUBLE SLEEPING THROUGH THE NIGHT					
5. I EAT AS MUCH AS I USED TO					
6. I ENJOY LOOKING AT, TALKING TO AND BEING WITH ATTRACTIVE WOMEN/MEN					
7. I NOTICE THAT I AM LOSING WEIGHT					
8. I HAVE TROUBLE WITH CONSTIPATION					
9. MY HEART BEATS FASTER THAN USUAL					
10. I GET TIRED FOR NO REASON					
11. MY MIND IS AS CLEAR AS IT USED TO BE					
12. I FIND IT EASY TO DO THE THINGS I USED TO					
13. I AM RESTLESS AND CAN'T KEEP STILL					
14. I FEEL HOPEFUL ABOUT THE FUTURE					
15. I AM MORE IRRITABLE THAN USUAL					
16. I FIND IT EASY TO MAKE DECISIONS					
17. I FEEL THAT I AM USEFUL AND NEEDED					
18. MY LIFE IS PRETTY FULL					
19. I FEEL THAT OTHERS WOULD BE BETTER OFF IF I WERE DEAD					
20. I STILL ENJOY THE THINGS I USED TO DO					

SDS RAW SCORE

SDS INDEX

Headache Questionaire

Name: .... age .... .sex ....
address: ....
....
vocation .... single/married
children .... vocation of spouse ....

BEGINNING of headache .... when ....

CIRMCUMSTANCES ....
associated factors: .... physical .... psychic ....

possible causes:
Position of head (trunk) .... / .... especially ....
weather.....(changes).....; warmth.....cold.....; get up.....; reclining.....; wake up.....; sleep.....;
light.....; to see.....; noise.....; hear.....; hunger.....; thirst.....; eating.....; drinking.....; food stuff:
alcohol.....; nicotine.....; coffeine.....; tiredness.....; menstruation.....; coitus.....;
other causes: ....

PRODROMES: yes.....; no.....; how long before the headache.......; duration.....;
symptoms: visual.....; auditory.....; emotional.....; shivering.....; pallor of face.....; light.....;
noise.....; micturition.....

LOCALISATION: type 1: front.....; temp.....; occip.....; irradiation? .....;
type 2: front.....; temp.....; occip.....; bilat.....; irradiation? .....;
irritation .... always? yes.....; no.....;

SPECIAL CHARACTERISTICS: equal.....; changing.....; how fast.....; slow.....; end:
fast.....; slow.....; intern sensitive scalp.....; fits? .....;
other ....

FREQUENCY: daily.....; monthly.....; .....times in a year; regul.....; lucid interval....., day.....,
month.....; longest lucid intervall: .....with therapy.....; occurs: day.....; night.... ac.... pc.... all the
time.....; seldom.....; on getting up mornings.....; on falling asleep.....; other: .....;

DURATION: seconds.....; min.....; hours.....; days.....; all the time.....; .....times per week
repeating: seldom.....; often.....; getting: worse.....; better.....; more often.....; more seldom.....;
unil.....; bilat.....

EFFECT OF: position.....; motion.....; pressure.....; bending forward.....; lie down.....;
warmth.....; cold.....; noise.....; darkness.....; light.....; eating.....; drinking.....; stools.....;
other.......(+ = better, - = worse)

SIMULTANEOUSLY WITH: tears.....; conjunctivitis.....; red eyes.....; edema of lid.....;
diplopia.....; scotoma.....; flickering.....; photophobia.....; blurred vision.....; other.......;
tinnitus .....; vertigo.....; clogged ears.....; nausea.....; vomiting.....; bad taste.....; bad smell.....;
cold hands.....; cold feets.....; sweating.....; fright.....; depression.....; disturbances of GI-system.....;
GU-system.....; speaking.....; hearing.....; motor performance.....; sensory performance.....; sensory
functions.....; local....or general....edema

END of headache: suddenly____; after sleep____; sweating____; eating____; micturition____; def.____; staying in darkness____, gradually____;

MIGRAINE EQU.: abdominal____; ophthalm____; precord____; other____;

HEREDITARY: father____; mother____; siblings____; other____ had: headaches____; epil.____; art.hypertension____; mental/psychiatr. diseases____; (which____); other____

ANAMNESIS WITH SICK PERSON: pregnancy____; birth____; accident____; fall____;(+ = path., - = no) diseases of childhood: meningitis____; enceph.____; pertussis____; measles____; scarlatina____; unknown syncopes____; enuresis____; neurol. or psychiatr. diseases____; other_____; surgical diseases_____; non surg. dis._____;
amplification with **RELATIVES:** ....

alcohol: yes____; no____; which____;
nicotine: yes____; no____; which____;

sex. anamnesis: onan.____; 1.intercourse__ (with age___); last interc.____;
menstruation anamn.: first____; every____days/duration of__days;
correlates with headache: yes____; no____; the day before____; after____; last menstr.____; children____; pregnancies____; abortion____;

SYSTEM REVIEW: card.vasc.____; G.I.____; G.U.____; Skin____; respirat.____; bone____; metabolic diseases____; nervous____; mental diseases____;

REPORTS up to now: ....
....

TREATMENT so far: ....
....

weight_____lbs; heigth_____in

DIAGNOSTIC IMPRESSION:

proposed therapy:

examination to be done:

Beginning of therapy: effect:

FINAL DIAGNOSIS: date

                                        signature of physician:

Questionaire for Symptoms of Migraine

---

aura before the attack

1) depressive mood
2) pressure on chest
   epigastric pain
     gastric pain
     belly-ache
     shivering
     cold fingers
     cold feet
3) urinary urgency
   stool urgency
   ravenous appetite
   thirst
   desire for salty food
   desire for sour food
   nightmares
4) hallucinations
     of smell
     of taste
5) hardness of hearing
   vertigo
   temporary disturbances of
   speech
   sensation
   vision

The Attack itself consists of:

1) headaches
   occipital
   forehead / eyes
   temple
   unilateral (one side only)
   wave-like
   sensitive to pressure on N.supraorbitalis
2) a) stomach
     nausea (beginning when ......)
     vomiting (ending when ......)
     only stomach symptoms
     upper abdomen right
     upper abdomen left
     around navel
     lack of appetite
     disgust of food stuff (which......)
   b) depressive mood
     apathetic state
     decreasing will power
     other psychic disorder
     pallor of face
     temporal artery pulsating strongly
     increased salivation
     wide lit slit
     wide pupil (on side of ache,may be?)

c) reddened face
   reddened conjunctivae
   narrow lid slit
   small pupil
   more tears
   upper lid drooping
d) swellings of hand
     of head
     of lids
     of lips
     of nose
     of mucous membranes of nose
     running nose
     sweating of face
e) cardiac aches, unspecified
   palpitations
   fast pulse
   trepidation
   feeling of anxiety
   only anxiety, nothing else
f) diarrhoea
   pain under left rib margin
   colitis
   urinary urgency
g) cold hands
   cold feet
   bursts of sweating
   numbness of skin (e.g.fingers etc.)
   fingertips become red
   fingers turn white
   swelling of female breasts
   secretion from breasts (which? ......)
   increased temperature
   attacks of pain (where? ......)
   attack of asthma
3) a) frontal pain
   upper jaw painful
   lower jaw painful
   yawn
   sneeze
   vertigo   buzzing ears?
   tingling ears (tinnitus)
   facial tic, uni- or bilateral
   oversensitivity of organs of
     vision
     smelling
     hearing
     tasting

b) visual disturbances, simple
  visual disturbances, scintillating
  sensory disturbance, unilateral
  speech disturbance
  lack of part of visual field (scotomata)
    simple
    combined with contralateral
    sensory disturbances
3) c) disturbed eye movements to nose
  disturbed eye movements to temple
  vomiting
  double vision
  drooping upper lid
  loss of hearing, right left
  loss of taste
d) vertigo
  disturbed equilibrium
  both these symtoms alone: yes no
e) other attacks, such as epilepsy
  patient
  relatives
f) migraine of relatives:
  mother
  father
  grandmother (of m, f)
  grandfather (of m, f)
  siblings
  children
  attacks during pregnancy:
    □ yes; □ no

Summary for the physician:

kind and number of symptoms:

AURA:

1) psychic
2) somatic
3) rare
4) hallucinations

ATTACK:

1) headache
2) accompanying symptoms:
  a) stomach
  b) psychic
  c) sympathetic:
      stimulation
      inhibition
  d) other sympathetic
  e) heart
  f) gut
  g) vessels
3)
  a) cranial nerves
  b) eyes
  c) ophthalm. migr.
  d) vertigo
  e) epilepsy
  f) migraine

not for the patient

Blood Tests in Patients Sensing
Changes in Weather

Test	results	
	the day without complaints	with most severe signs
BSR		
leucocytes		
lymphocytes %		
eosinophil leucocytes %		
young leucocytes		
iron		
copper		
PTT in %		
Rheuma serology		
ASLO		
Latex		
CRP		
Waaler-Rose		
Pherogram		
Albumin fraction		
α_1		
α_2		
β		
γ-globulin		
Albumin/globulin ratio		
Bio – Weather		

O_2

Observation on Weather Factors

Factors	on day before	on day of	on day after
		c o m p l a i n t s	
atmospheric pressure			
air temperature			
irradiation			
gaseous pressure			
humidity of atmosphere			
evaporation			
temperature of soil			
clouds			
sun shine			
wind strength			
velocity of winds			
precipitation			
snow			
DATE			

REFERENCES

[1]   ADAMS, C. B. T. : Microvascular Compression: An Alternative View and Hypothesis. J. Neurosurgery **70**: 11-12. 1989. (1/3 failures, therefore therapy unconvincing).

[2]   ALMAY, B. G. : Patients with Idiopathic Pain Syndromes. A Clinical, Biochemical and Neuroendocrinological Study. Umeå University Medical Dissertations Series No. 191, Umeå, 1987.

[3]   ALMAY, B. G. , F. JOHANSSON, L. V. KNORRING, T. SAKURADA and L. TERENIUS: Long-term High Frequency Transcutaneous Electrical Nerve Stimulation (hi.TNS) in Chronic Pain; Clinical Response and Effects on CSF-Endorphins, Monoamine Metabolites, Substance P-like Immunoreactivity (SPLI) and Pain Measures. J. Psychosomatic Research **29**: 247-257, 1985.

[4]   ALTHAUS, J. : Über elektrische und elektrochemische Anaesthesie. Wien. med. Woschr. **9**: 433-435, 1859.

[5]   AMMER, K. und F.L.JENKNER: Vergleichende Untersuchungen zur Wirkung verschiedener Behandlungsmethoden bei Periarthropathia humero-scapularis unter besonderer Berücksichtigung der elektrischen Nervenblockade. In: BERGER, M., G. GERSTENBRAND und K. LEWITT. Schmerz und Bewegungssystem. Fischer Verlag, Stuttgart, New York (Schmerzstudien Bd. 6) 293-300, 1984.

[6]   ARING, C. D. : The Early Treatment of Stroke. Cincinnati J. Med. **30**: 304ff., 1949.

[7]   ATEFIE, K. and F. L. JENKNER: Thalamic Pain. Schmerz-Pain-Douleur, 9(3a): 221-227, 1988.

[8]   BAROLIN, G. S. : Atypische Migränen. Wien, klin. Woschr. **75**: 293-301, 1963.

[9]   BAROLIN, G. S. und M. MEIXNER: Vertebragen (mit-)verursachter Kopfschmerz. Therapiewoche **31**: 6987-7008, 1981.

[10]   BARRON, D. H. and B. A. C. MATTHEWS: Intermittent Conduction in the Spinal Cord. J. Physiol. (London)**85**: 73-103, 1938.

[11]   BAYINDIR, D. ,T. PAKER, B. AKPINAR, S. ERENTURK, D. ASKIN and A. AYTAC: Use of transcutaneous nerve stimulation in the control of postoperative chest pain after cardiac surgery. J. Cardiothorac. Vasc. Anaesth. **5**: 589- 591, 1991.

[12]   BECKER, W.: Der periartikuläre Kopfschmerz. Therapiewoche **33**: 3797-3801, 1983.

[13]   BERGSMANN, O. und O. BISCHKO: Methoden der Therapie des Brustschmerzes: Akupunktur bei Thoraxschmerzen. Therapiewoche **32**: 2459-2462, 1982.

[14]   BERNHARD, P. : Chronische funktionelle Schmerzzustände des Bewegungsapparates aus psychosomatischer Sicht. Therapiewoche **33**: 3837-3843, 1983.

[15]   BETTENDORF, G., M. BRECKWOLDT, J. HAMMERSTEIN, P. J. KELLER, H. KUHL und B. RUNNEBAUM: "Züricher Empfehlungen zur oralen Kontrazeption". Geburtshilfe und Frauenheilkunde **46**: VIII, 1986.

[16]   BIDWAL, A. V., C. R. ROGERS, M. PEARCE and T. H. STANLEY: Postoperative Stellate Ganglion Blockade to Prevent Hypertension Following Coronary Artery Operations. Anesthesiology **51**: 345-347, 1979.

[17]   BIRKMAYER, W.: Der Wandel der vegetativen Dystonie in den letzten zwei Jahrzehnten, Wiener klin. Woschr. **81**: 570-571, 1969.

[18]   BIRKMAYER, W. und P. RIEDERER: Neurotransmitter und menschliches Verhalten. Springer-Verlag Wien, New York 1986. Besonders S. 32-43.

[19]   BLOEDEL, J. R. and D. B. MCCREERY: Organisation of Peripheral and Central Pain Pathways. Minneapolis Pain Seminar, December 6-8, 1973.

[20]   BONICA, J. J. : The Management of Pain. Lea and Febiger, Philadelphia, PA., 1954.

[21]   BONICA, J. J. : Editorial: Chronischer Schmerz als Krankheit. Triangel Sandoz **20**: 1-6, 1981.

[22]   BONICA, J. J. : Physiologische Mechanismen von Schmerz und Schmerztherapie. Triangel Sandoz **20**: 7-18, 1981.

[23]   BRAUN, A. : Differentialindikation zur Schmerztherapie durch Akupunktur. Therapiewoche **33**: 3820-3822, 1983.

[24] BREITBACH, A., MÜSCH, H.: Selektive Reizung vegetativer Nervenfasern im N. ischiadicus des Frosches. Pflüg. Arch. ges. Physiol. **241**: 360-369, 1938.

[25] BREMERICH, A. und KRISCHEK-BREMERICH. P.. Therapie der Trigeminusneuralgie. Ein Überblick. Dtsch. Z. Mund-Kiefer-Gesichtschir. **15**: 369-375, 1991.

[26] BRENA, S. F. and S. J. CHAPMAN: Validity of the Emory Pain Estimate Model. Anesthesiology Review **IX**: 42-45, 1982.

[27] BREZOWSKY, H.: Physiologische und pathophysiologische Abläufe beim Menschen in verschiedenen Klimagebieten Bayerns. Münch. Med. Wochenschr. **102** (Heft 51): 2533-2538, 1960.

[28] BRIX, E., F. L. JENKNER: Unveröffentlichte Beobachtungen über evozierte Potentiale am Computer-EEG. 1976.

[29] BROMM, B. und E. SCHAREIN: Methoden zur Quantifizierung der klinischen Wirksamkeit von Analgetica. Therapiewoche **34**: 5207-5214, 1984.

[30] BÜHRING, M. und P. Kirchner: Methoden der Therapie des Brustschmerzes: Physikalische Therapie. Therapiewoche **32**: 2465-2469, 1982.

[31] BURTON, C.: Dorsal Column Stimulation: Optimization of Application. Minneapolis Pain Seminar Dec. 6-8, 1973. Surg. Neurol.**4**: 171-179, 1975.

[32] CAIN, R. C. : How a Dallas Doctor Controls the Aches and Pains. International Management, May 1983, p. 64.

[33] CAMPBELL, J. N. : Examination of Possible Mechanisms by Which Stimulation of the Spinal Cord in Man Relieves Pain. Appl. Neurophysiol. **44**: 181-186, 1981.

[34] CAMPBELL, J. N. and D. M. LONG: Peripheral Nerve Stimulation in the Treatment of Intractable Pain. J. Neurosurg. **45**: 692-699, 1976.

[35] CAMPBELL, J. N. and D. M. LONG: Transcutaneous Electrical Stimulation for Pain: Efficacy and Mechanism of Action. Diagnosis and Treatment of Chronic Pain (N. H. Hendler, ed.) John Wright - PSG Inc. Boston, Bristol, London, 1982.

[36] CAMPBELL, J. N. and R. H. LAMOTTE: Latency to Detection of First Pain. Brain Research **266**: 203-208, 1983.

[37] CAMPBELL, J. N. and R. A. MEYER: Sensitization of Unmyelinated Nociceptive Afferents in Monkey Varies with Skin Type. Journal of Neurophysiology **49**: 98-110, 1983.

[38] CAMPBELL, J. N. , TAUB, A.: Local analgesia from percutaneous electrical stimulation. A peripheral mechanism. Arch. Neurol. (Chicago) **28**: 347-350, 1973.

[39] CARMICHAEL, J. K.: Treatment of herpes zoster and postherpetic neuralgia. Am. Fam. Physician **44**: 203-210, 1991.

[40] CAUTHEN, J. C. and E. RENNER: Transcutaneous and Peripheral nerve Stimulation for Chronic Pain States. Surg. Neurol.**4**: 102-104, 1975.

[41] CHAN-LIAO, M. , J. LIAO, Y. W. WU and C. S. LIAO: Transcutaneous electric nerve stimulation for the treamtment of abdominal zoster paresis. Ma-Tsui Hsuch Te Chi **29**: 559-563, 1991.

[42] CHARTERS, T, H.: Electrode Systems and Therapeutic Currents for Nerve Block and Traditional TENS. Schmerz-Pain-Douleur, 9 (3a): 146-157, 1988.

[43] CLARK, K.: Electrical Stimulation of the Nervous System for Control of Pain: University of Texas South-Western-Medical School Experience. Surg. Neurol. **4**: 164-166, 1975.

[44] CRUE, B. L., jr.: Pain Research and Treatment. Academic Press, New York, San Francisco, London, 1975.

[45] CZISKE, R. und B. FISCHER: Wie zuverlässig sind Schmerzangaben aus der Erinnerung des Patienten? Therapiewoche **35**: 282-286, 1985.

[46] COLLINS, W. F. and C. T. RANDT: Evoked Central Nervous System Activity Related to Peripheral Unmyelinated or "C"-fibers in Cat. J. Neurophysiol. **21**: 345-352, 1958.

[47] COLLINS, W. F. and C. T. RANDT: Midbrain Evoked Responses Relating to Peripheral Unmyelinated or "C"-fibers in Cat. J. Neurophysiol. **23**: 47-53, 1960.

[48] DALESSIO, D. J.: Some Current Data on Headache Research. Triangel Sandoz **20**: 33-42, 1981.

[49] DANDY, W. E.: Concerning the Cause of Trigeminal Neuralgia. Am. J. Surg. **24**: 447-455, 1934 (p. 450 Fig. 3 shows "Indentation by superior cerebellar artery"; in 66 cases or 30,7 %).

[50] DAVIS, R. and R. LENTINI: Transcutaneous nerve Stimulation for Treatment of Pain in Patients with Spinal Cord Injury. Surg. Neurol. **4**: 100-101, 1975.

[51] DE FELICE, E. A. and A. SUNSHINE: Basic Principles in the Management of Pain. Triangel Sandoz **20**: 43-48, 1981.

[52] DE MAR, E. A.: The Use of Transcutaneus Nerve Block on Traumatized Muscle. Schmerz, Pain, Doleur **9** (3a): 201-203, 1988.

[53] DEUTSCH, E. und G. GEYER: Laboratoriumsdiagnostik (2. Aufl.). Verlag Brüder Hartmann, Berlin, 1975.

[54] DICKEL, H. A. , H. H. DIXON, R. A. COEN and R. D. PETERSON: Fatigue. Northwest Medicine **51**: 32-35, 1952.

[55] DICKEL, H. A. and H. H. DIXON: The Fatigue Syndrome. The Amer. J. Nursing **53** (6):1-3, 1953.

[56] DIRNAGL, K. und J. KUGLER: Wetter und Kopfschmerz. Therapiewoche **30**: 8066-8076, 1980.

[57] DIXON, H. H. , H. A. DAVENPORT and S. W. RANSON: Chemical Studies of Muscle Contracture. J. Biol. Chem. **82**: 61-70, 1929.

[58] DIXON, H. H. and H. A. DICKEL: Tension Headache. Northwest Medicine **66**: 817-820, 1967.

[59] EBERSOLD, M. J., E. R. LAWS, H. H. STONNINGTON and G. K. STILLWELL: Transcutaneous Electrical Stimulation for Treatment of Chronic Pain; A Preliminary Report. Surg. Neurol, **4**: 96-99, 1975.

[60] ECCLES, J. C.,: Facing Reality, Heidelberg Science Libery. Vol. 13, Springer Verlag, New York, Heidelberg, Berlin; 1950.

[61] EDEL, H.: Fibel der Elektroagnostik und Elektrotherapie. 4. Aufl., Steinkopf, Dresden, 1977.

[62] ENZELSBERGER, H. , W. D. SKODLER und E. KUBISTA: Zur Verbesserung der Dopplersonographiebefunde nach transkutaner Elektrostimulation bei Frauen mit Plazentainsuffizienz. Z. Geburtshilfe Perinatol. **195**: 172-175, 1991.

[63] ERLANGER, J. and H. S. GASSER: Electric Signs of Nervous Activity. Univ. of Pennsylvania Press 1937.

[64] FANCHAMPS, A.: Einteilung, Pathogenese und medikamentöse Behandlung der Migräne und verwandter Kopfschmerzen. Med. Welt **26**: 1518-1523, 1975.

[65] FASSL, A. and F. L. JENKNER: Electroblock of Superior Laryngeal Nerve for Relief of Bronchial Obstruction. Schmerz-Pain-Douleur 9(3a): 249-254, 1988.

[66] FIELDS, H. L. and J. D. LEVINE: Pain-Mechanics and Management. West J. Med., **141**: 347-357, 1984.

[67] FLORA, G. and K. SCHWAMMBERGER: Behandlung der Hyperhidrose durch Sympathikotomie. Paracelsus Arch. d. prakt. Medizin **9**: 239-240, 1972.

[68] FLÖTER, TH.: Methoden der Therapie des Brustschmerzes: Behandlung mit Lokalanaesthesie. Therapiewoche **32**: 2444-2453, 1982.

[69] FLY H. .: Shocking therapy: uses of transcutaneous electric nerve stimulation in dermatology. Dermatol. Clin. **9**: 189-197, 1991.

[70] FOX, E. J. and R. MELZACK: Transcutaneous Electrical Stimulation and Acupuncture: Comparison of Treatment for Low-back Pain. Pain **2**: 141-195, 1976.

[71] FRANCIS, J. B.: Extracting Teeth by Galvanism. Dent. Dep. **1**: 65-69, 1858.

[72] GAILLARD, L. C. in KELLAWAY, P.: The William Osler Medal Essay: The Part Played by Electric Fish in the Early History of Bioelecticity and Electrotherapy. Bull. Hist. Med. (Baltimore) **XX**: 112-137, 1946.

[73] GEDDES, L. A. and A. SIMMONS: Artificial respiration in the dog by percutaneous bilateral phrenic nerve stimulation. Amer. J. Emerg. Med. **90**: 527-529, 1991.

[74] GEIGER, TH.: Methoden der Therapie des Brustschmerzes: Manuelle Therapie. Therapiewoche **32**: 2454-2456, 1982.

[75] GEORGI, P.: Wert der nuklearmedizinischen Diagnostik bei artikulären und periar- tikulären Schmerzzuständen. Therapiewoche 33: 3833-3834, 1983.

[76] GERBERSHAGEN, H. U.: Bemerkungen zur Schmerzdiagnostik und medikamentösen Therapie. Therapiewoche 34: 5203-5207, 1984.

[77] GFELLER, F.: Untersuchungen über die allgemeinen physiologischen Eigenschaften des Sympathicus geprüft am Nervus accelerans des Frosches. Z. Biol. 89: 202-216, 1929.

[78] GRANAT, M. , J. F. KEATING, A. C. SMITH, M. DELARGY and B. J. ANDREWS: The use of functional electric stimulation to assist gait in patients with incomplete spinal cord injury. Disabil. Rehabil. 14: 73-77,1992.

[79] GREGOR, M. and M. ZIMMERMANN: Dorsal Root Potentials Produced by Afferent Volleys in Cutaneous Group III Fibres. J. Physiol. (London), 232: 413, 1973.

[80] GROSS, D.: Brustschmerz: Pathogenese – Klinik – Therapie. I. Klinisches Bild, II. Faktorenanalyse. Therapiewoche 32: 2388-2396, 2398-2410, 1982.

[81] GROSS, D.: Schmerzkonferenz – Schmerzklinik – Schmerzambulanz. Therapiewoche 32: 2490, 1982.

[82] GROSS, D.: Sympathalgien des Nacken-Schulter-Arm-Bereichs. Therapiewoche 32: 2424-2432, 1982.

[83] GROSS, D. und D. LANGEN: Therapie über das Nervensystem. Bd. 8. Schmerz und Schmerztherapie. Hippokrates Verlag Stuttgart, 1971.

[84] GRUNERT V., G. PENDL und M. SUNDER-PLASSMANN: Ergebnisse und Komplikationen der operativen Behandlung der Trigeminusneuralgie. Neurochirurgie 14: 127-133, 1980.

[85] GUELRUD, M. , A. ROSSITER, P. F. SOUNEY and M. SULBARAN: Trancutaneous electric nerve stimulation decreases lower esohageal spincter presure in patients with achalasia. Dig. Dis. Sci. 36: 1029 - 1033, 1991.

[86] GUILLAUME, J., S. DE SEZE et G. MAZARS: Chirurgie cerebrospinale de la douleur. Presse universitaire de France, Paris 1944.

[87] HALPERN, L. M.: Analgesic Drugs in the Management of Pain. Symposium on Pain: Part II. Arch. Surg. 112: 861-869, 1977.

[88] HAN, H. S. , X. H. CHEN, S. L. SUN, X. J. XU, Y. YUAN, S. C. YAN, J. X. HAO and L. TERENIUS: Effect of low- and high-frequency TENS on Met-enkephalin Arg-Phe and Dynorphin A immunoreactivity in human lumbar CSF. Pain 47: 295-298, 1991.

[89] HANNIGTON-KIFF, J. G.: Pharmacological Target Blocks in Painful Dystrophic Limbs. In: WALL, P. O. and R. MELZACK (eds.) Textbook of Pain. Edinburgh, Churchill- Livingstone 1989: 754-66.

[90] HARDDY, J. D., L. W. FABIAN and M. D. TURNER: Electrical Anastesia for Major Surgery. Journ. Am. Med. Ass. 175: 599-600, 1961.

[91] HART, F. D.,(ed): The Treatment of Chronic Pain. MTP, London, 1974.

[92] HAUGER, R.: Schmerzbehandlung mittels TNS. Studio M Magazin (Marienhospital Aachen-Burtscheid) 6: 14-16. 1985

[93] HAUPTMANN, P. J. : Electrocardiographic artifact with a transcutaneous electrical nerve stimulator. Int. J. Cardiol. 34: 110-112, 1992.

[94] HEYCK, H.: Neue Beiträge zur Klinik und Pathogenese der Migräne. G. Thieme Verlag, Stuttgart, 1956.

[95] HOFFMANN, S. D. und O. EGLE: Zum Beitrag von J. J. GROEN über das psychogene Schmerzsyndrom – zugleich ein Plädoyer für die Erweiterung des Konversionsbegriffes. Psychother. med. Psychol. 34: 25-26, 1984.

[96] HOLLMEN, A. und SAUKKONEN: Zur postoperativen Schmerzausschaltung nach Ober- bauchoperationen. Anaestesist 180: 289, 1969.

[97] HOPF, H. C.: Brustschmerz und Nervensystem: Rückenmark, Wurzel, periphere Ner- ven. Therapiewoche 32: 2420-2423, 1982.

[98] HOPPENSTEIN, R.: Electrical Stimulation of the Ventral and Dorsal Columns of the Spinal Cord for Relief of Chronic Intractable Pain: Preliminary Report. Surg. Neurol. 4: 187-194, 1975.

[99] HOPPENSTEIN, R.: Precutaneous Implantation of Chronic Spinal Cord Electrodes for Control of Intractable Pain: Preliminary Report. Surg. Neurol. 4: 195-198, 1975.

221

[100] HOSOBUCHI, Y., J. E. ADAMS and B. RUTKIN: Chronic Thalamic and Internal Capsule Stimulation for the Control of Central Pain. Surg. Neurol. 4: 91-95, 1975.

[101] HÜBNER, B.: Radiologische Diagnostik von Spinalnervenwurzelsyndromen mit Hilfe von Kontrastmitteln. Therapiewoche 27: 7219-7223, 1977.

[102] HUGHES, J., T. W. SMITH, H. W. KOSTERLITZ, L. A. FOTHERGILL, B. A. MORGAN and H. R. MORRIS: Identification of Two Related Pentapeptides from the Brain with Potent Opiate Agonist Action. Nature, 258: 577-579, 1975.

[103] HUNSTEIN, W.: Labordiagnostik inklusive Synoviaanalyse bei artikulären und periartikulären Schmerzzuständen. Therapiewoche 33: 3811-3819, 1983.

[104] HUNT, W. E., J. H. GOODMAN and W. G. BINGHAM: Stimulation of the Dorsal Spinal Cord for Treatment of Intractable Pain: Preliminary Report. Surg. Neurol. 4: 153-156, 1975.

[105] IGNELZI, R. J. and J. K. NYQUIST: Direct Effect of Electrical Stimulation on Peripheral Nerve Evoked Activity: Implications in Pain Relief. J. Neurosurg. 45: 159-165, 1976.

[106] IGNELZI, R. J. and J. K. NYQUIST: Excitability Changes in Peripheral Nerve Fibers After Repetitive Electrical Stimulation: Implications in Pain Modulation. J. Neurosurg. 51: 824-833, 1979.

[107] INMAN, V. T. and J. B. DE C. M. SAUNDERS: Referred Pain from Skeletal Structures. J. Nerv. Ment. Dis. 99: 660, 1944.

[108] JANETTA, P. J.: Observations on Etiology, Hemifacial Spasm, Acoustic Nerve Dysfunction and Glossopharyngeal Neuralgia. Neurochirugia (Stuttgart) 20: 145-54, 1977.

[109] JANTSCH, H. und F. SCHUHFRIED: Niederfrequente Ströme zur Diagnostik und Therapie. W. M. Maudrich, München-Bern, 1974.

[110] JENKNER, F. L.: Rheoencephalography. C. C. THOMAS, Springfield, Ill. 1962. Unauthorisierte Übersetzung bei Medizina, Moskau, 1966. Clinical Rheoencephalograhy. Eigenverlag Wien 1986.

[111] JENKNER, F. L.: A New Multi-purpose Electrode Material for General Biomedical Application. The Bulletin of the DOW Corning Center for Aid to Medical Research 9: 10, 1967.

[112] JENKNER, F. L.: Rheoencephalographic Differentiation of Vascular Headaches of Varying Causes. Ann. New York Acad. Sci. 170: 661-666, 1970.

[113] JENKNER, F. L.: Möglichkeiten und Besonderheiten chirurgischer Schmerzausschaltung. Acta chir. Austr. 5: 123-129, 1973.

[114] JENKNER, F. L.: Schmerzbehandlung durch transdermale Reizstrombehandlung. Wien. klin. Woschr. 89: 125-131, 1977.

[115] JENKNER, F. L.: Rheoencephalographische Untersuchungen zum Wirkungsmechanismus der Stellatumblockade. Folia angiol. XXV: 47-52, 1977.

[116] JENKNER, F. L.: Impulsgalvanisation des Ganglion Stellatum. Elektrische Stellatumblockade. Wien. med. Woschr. 127: 59-62, 1977.

[117] JENKNER, F. L.: Die elektrische Blockade von sympathischen und somatischen Nerven von der Haut aus. Wien. klin. Woschr. 92: 233-240, 1980.

[118] JENKNER, F. L.: Nervenblockaden auf pharmakologischem und auf elektrischem Weg. Indikationen und Technik. 4. Aufl., Springer, Wien-New York, 1983.

[119] JENKNER, F. L.: Das Cervikalsyndrom. Springer, Wien-New York, 1982.

[120] JENKNER, F. L.: Electric Nerve Block. Springer, Wien-New York, 1986.

[121] JENKNER, F. L.: Treatment of Cervical Syndromes. In: J. J. GERHARDT, W. RAINER, B. O. SCHWEIGER und P. P. KING (Ed.) Interdisciplinary Rehabilitation in Trauma. Williams and Wilkins, Baltimore-London-Los Angeles-Sidney, 1987: 540-550.

[122] JENKNER, F. L.: Wetterfühligkeit aus medizinischer Sicht. Handbuch für den Kurarzt. Verlag der österreichischen Ärztekammer Wien, 1988: 189-195; und: Wetterfühligkeit: was ist das? Was macht das? Symptomenverschiebungen bei 5000 Patienten mit Zervikalsyndrom. Ebenda 1989: 99-111.

[123] JENKNER, F. L.: TENS-Therapy of Trigeminal Neuralgia. Schmerz-Pain-Douleur 9(3a): 179-183, 1988.

[124] JENKNER, F. L.: Rectal and Bladder Sphincter Training for Incontiency. Schmerz-Pain-Douleur 9(3a): 255- 259, 1988.

222

[125] JENKNER, F. L.: Transdermale Schmerztherapie. In: REISIGL (Ed.) Schmerztherapie. Bibliomed 1991: 91-112 (vorgetragen 1989).

[126] JENKNER, F. L.: Verschiedene Arten der transdermalen elektrischen Nervenstimulation (TENS) in der Schmerztherapie. Hausarzt 1993(3): 14-17.

[127] JENKNER, F. L. and K. ATEFIE: Pain from bone athrophy (SUDECK's athrophy): A case report. Schmerz-Pain-Douleur 9(3a): 216-220, 1988.

[128] JENKNER, F. L. and A. FASSL: Unilateral Phrenic Nerve Palsy: Management by Transcutaneous Phrenic Nerve Stimulation. Schmerz-Pain-Douleur 9(3a): 259-262, 1988.

[129] JENKNER, F. L. und E. KIESEWETTER: Über die klinische Wertigkeit eines kurzzeitigen Karotiskompressionstests. Folia Angiol. XXIV: 153-159, l978.

[130] JENKNER, F. L. and F. SCHUHFRIED: Transdermal Transcutaneous Electric Nerve Stimulation for Pain: The Search for an Optimal Waveform. Appl. Neurophysiol. 44: 330-337, 1981.

[131] JENKNER, K. W. and JENKNER, F. L.: A modern view on a famous roman physician and writer's report. Schmerz-Pain-Douleur 9(3a): 132-140. 1988.

[132] JOHNSON, M. I. , C. H. ASHTON and J. W. THOMPSON: An in-depth study of transcutaneous electric nerve stimulation (TENS). Implications for clinical use of TENS. Pain 44: 221-229, 1991.

[133] JOHNSON, M. I. , C. H. ASHTON and J. W. THOMPSON: The consistency of pulse frequencies and pulse patterns of transcutaneous electrical nerve stimulation (TENS) used in chronic pain patients. Pain 44: 231-234, 1991.

[134] JOHNSON, M. I. , C. H. ASHTON, D. R. BOUSFIELD and J. W. THOMPSON: Analgesic effect of different pulse patterns of transcutaneous electric nerve stimulation on cold induced pain in normal subjects. J. Psychosom. Res. 35: 313-321,1991.

[135] KAADA, B. , E. FLATHEIM and L. WOIE: Low- frequency transcutaneous nerve stimulation in mild/moderate hypertension. Clin. Physiol. 11: 161-168, 1991.

[136] KAINDL, F., K. POLZER und F. SCHUHFRIED: Rheographie. Steinkopf, Darmstadt. 1967.

[137] KANE, K., TAUB, A.: A history of local electric analgesia. Pain 1: 125-138, 1975.

[138] KEIDEL, W. D.: Informationsverarbeitung. In: Kurzgefaßtes Lehrbuch der Physiologie. Herausg.: W. D. Keidel, 1. Aufl. Stuttgart, G. Thieme, 1967.

[139] KEIDEL, W. D.: Sinnesphysiologie: Teil I. Allgemeine Sinnesphysiologie. Visuelles System. 2. Aufl. Springer, Berlin-Heidelberg-New York, 1976.

[140] KERR, F. W. L.: Pain. A central inhibitory balance theory. Mayo Clinic. Proc. 50: 685-690, 1975.

[141] KLEIN, K.: Hirngefäßerkrankungen und Autoregulation. Dr. Thiemann Gmbh, Lünen, 1976.

[142] KLINGENBURG, M., E. GAUS und R. WÖRZ: Psychogener und psychotisch bedingter Brustschmerz. Therapiewoche 32: 2435-2438, 1982.

[143] KNORRING, L. V., B. G. L. ALMAY, F. JOHANSSON, G. SCHUBER, L. TERENIUS: Changes in CSF endorphines and monoamine metabolites related to treatment with high frequency transcutaneous electrical nerve stimulation. Nord. Psyk. Tidsk. 39(Suppl. 11): 83-90, 1985.

[144] KNORRING, L. V., B. G, L. ALMAY, F. JOHANSSON, L. TERENIUS: Pain perception and endorphin levels in cerebrospinal fluid. Pain 5: 359-365, 1978.

[145] KNORRING, L. V., F. JOHANSSON, B. G, L. ALMAY: The importance of the endorphin systems in chronic pain patients. In: Endorphins and Opiate Antagonists in Psychiatric Research (Shan and Donald, ed.), pp. 407-426. Plenum Publishing Corporation. 1982.

[146] KRÄMER, J.: Lumbalgie: Spritzen Sie auch so? Med. Tribune Nr. 11: 62, 1982.

[147] KRAINICK, J. U., U. THODEN and T. RIECHERT: Spinal Cord Stimulation in Post-amputation Pain. Surg. Neurol. 4: 167-l70, 1975.

[148] KRAUS, H.: Clinical Treatment of Back and Neck Pain (An Exercise Manual). McGraw-Hill. New York, 1970.

[149] KRAUTHAMMER, V.: Modulation of Conduction at Points of Axonal Bifurcation by Applied Electric Fields. IEEE Transactions on Biomedical Engineering. **37**: 515-519, 1990.

[150] KÜGLER, H.: Medizin-Meteorologie nach den Wetterphasen. J. F. Lehmanns Verlag. München, 1975.

[151] KÜHNER, A.: Vom Plexus lumbalis, Plexus sacralis und Plexus brachialis ausgehende Schmerzzustände. Therapiewoche **33**: 3802-3810, 1983.

[152] KUBISTA, E., H. ENZELSBERGER und W. D. SKODLER: Der Einflußdes TNS auf sonographische Funktionsparameter bei Plazentainsuffizienz. Gynäkol. Rundsch. **31** (Suppl. 2): 154-155,1991.

[153] KUBISTA, E., H. KUCERA und P. RISS: Die Wirkung der transkutanen Nervenstimulation auf den Wehenschmerz. Geburtsh. u. Frauenheilk. **38**: 1079-1084, 1978.

[154] KUNERT, W.: Rheographische Messungen im Vertebralis-Stromgebiet. Der Nervenarzt **32**: 34-38, 1961.

[155] KUSCHINSKY, K.: Methoden der Therapie des Brustschmerzes: Medikamentöse Therapie. Therapiewoche **32**: 2441-2442. 1982.

[156] LANGBERG, G. J. : Ultra-low-frequency TENS: A Well- kept Secret. Pain Management 1990 (9/10): 278-280.

[157] LARBIG, W.: Kultur und Schmerz. Untersuchungen zur zentralnervösen Schmerzverarbeitung: Empirische Befunde und klinische Konsequenzen. Psychomed **1**: 17-26, 1989. Schmerz-Verlag Kohlhammer, Stuttgart 1982.

[158] LARSON, S. J., A. SANCES, J. F. CUSIK, G. A. MEYER and T. SWIONTEK: A Comparison Between Anterior and Posterior Spinal Implant Systems. Surg. Neurol. **4**: 180-186, 1975.

[159] LEA, P.: Delivering women from labor pain. Can. Nurse **88**: 17-19, 1992.

[160] LECHNER, W. , F. JARUSCH, E. SOLDER, A. WAITZ-PENZ und G. MITTERSCHIFFTHALER: Verhalten von Beta-Endorphin während der Geburt unter elektrischer Nervenstimulation. Zentralbl. Gynäkol. **113**: 439-448, 1991.

[161] LECHNER, H. und G. LADURNER: Zervikale und lumbosakrale Wurzelsyndrome aus neurologischer Sicht: Anatomische Grundlagen. Therapiewoche **27**: 7211-7218, 1977.

[162] LECHNER, H. und E. OTT: Proceedings of the Headache Classification Meeting in Graz, December 4th,1981. Neurologia et Psychiatria, 5, Suppl. 1: 1-64, 1982.

[163] LESKY, E.: Hundert Jahre Theorie und Therapie des Schädelhirntraumas. Wiener Med. Woschr, **42**: 711-714, 1969.

[164] LEVINE, D. Z.: Burning pain in an extremity. Breaking the destructive cycle of reflex sympathetic dystrophy. Postgrad Med. **90**: 175-178, 1991.

[165] LEWIT, K,: Kopfschmerz: Wirbelsäule als Auslöser. Manuel. Med.6: 62, 1968.

[166] LISSMANN, H. W. : On the function and evolution of electric organs in fish. J. Exp. Biol. **35**: 156-191, 1958.

[167] LOESER, J. D.: Dorsal Rhizotomy for the Relief of Chronic Pain. J. Neurosurg. **36**: 745-750, 1972.

[168] LONG, D. M.: Fifteen years of transcutaneous electric nerve stimulation for pain control. Stereotact. Funct. Neurosurg. **56**: 2-19, 1991.

[169] LONG, D. M. and D. E. ERICKSON: Stimulation of the Posterior Columns of the Spinal Cord for Relief of Intractable Pain. Surg. Neurol. **4**: 134-141, 1975.

[170] LUE, T. F., R. A. SCHMIDT and E. A. TANAGHO: Electrostimulation and Penile Erection. Urol. Int. **40**: 60-64, 1985.

[171] LÜBEN, V., H. F. HERGETH, M. KRAMER und D. PATSCHKE: Hämodynamische Untersuchung unter Naloxone während der Elektrostimulationsanalgesie (ESA) als Beitrag zur Endorphintheorie. Anästhesie, Intensivmedizin, Notfallmedizin **16**: 180-183, 1981.

[172] MACHALEK, A., H. TILSCHER, M. FRIEDRICH und E. POLT: Der Einflußdes Wetters auf den Verlauf von Lumbalsyndromen. Z. Orthop. **118**: 376-384, 1980.

[173] MANNHEIMER, C., C. A. CARLSSON, H. EMANUELSSON, A. VEDIN, F. WAAGSTEIN and C. WILHELMSSON: The Effects of Transcutaneous Electrical Nerve Stimulation in Patients with Severe Angina Pectoris. Circulation **71**: 3089-3106, 1985.

[174]   MANNHEIMER, J. S. and G. N. LAMPE (eds.): Clinical Transcutaneous Electrical Nerve
         Stimulation. F. A. Davis, Philadelphia, 1984.
[175]   MANNICHE, C. , E. LUNDBERG, I. CHRISTENSEN, L. BENTZEN and G. HESSELSŒ: Intensive
         dynamic back exercises for chronic low back pain: a clinical trial. Pain 47: 53-56,
         1991.
[176]   MARCHAND, S. , M. C. BUSHNALL and G. H. DUNCAN: Modulation of heat pain per-
         ception by high frequency transcutaneous electric nerve stimulation (TENS). Clin.
         J. Pain 7: 122-129, 1991.
[177]   MARBERGER, M., F. L. JENKNER: Transdermale Elektrotherapie bei Metastasenschmerz.
         Verhandlungsbericht der Deutschen Gesellschaft für Urologie, 34. Tagung, 20. bis 23.
         Oktober 1982, Hamburg. Springer, Berlin-Heidelberg-New York-Tokyo. 1983.
[178]   MARKOVICH, S. E.: Pain in the Head: A Neurological Appraisal. In: Gelb, H. (ed.):
         Clinical Management of Head, Neck and TMJ Pain and Dysfunction. A Multi-
         Disciplinary Approach to Diagnosis and Treatment. W. B. Saunders Co., Philadel-
         phia, London, Toronto, 1977.
[179]   MASON, S. F.: A History of the Sciences (p. 474 ff.) The Macmillan Company, New
         York, 1970.
[180]   MAZARS, G. J.. Intermittent Stimulation of Nucleus Ventralis Posterolateralis for
         Intractable Pain. Surg. Neurol. 4: 93-95, 1975.
[181]   MELZACK, R.: The Puzzle of Pain. Basic Books Inc., Harper Torch Books TB 5022,
         New York, 1973.
[182]   MELZACK, R. and P. D. WALL: Pain Mechanisms: A New Theory, Science 150: 971-
         979, 1965.
[183]   MELZACK, R., P. VETERE and L. FINCH: Transcutaneous Electrical Nerve Stimulation
         for Low Back Pain. Physical Therapy, 63: 489-493, 1983.
[184]   MEYER, G. A., H. L. FIELDS: Causalgia treated by selective large fiber stimulation of
         peripheral nerve. Brain 95: 163-168, 1972.
[185]   MEYER, R. A. and J. N. CAMPBELL: Myelinated Nociceptive Afferents Account for the
         Hyperalgesia That Follows a Burn to the Hand. Science 213: 1527-29, 1981.
[186]   MEYER, R. A. and J. N. CAMPBELL: Peripheral Neural Coding of Pain Sensation.
         Johns Hopkins APL, Technical Digest 2: 164-171, 1981.
[187]   MILTNER, F. O.: Indikationen zur Langzeitstimulation der dorsalen Rückenmarks-
         strukturen in der Therapie chronischer Schmerzen. Medtronic, p. 1-8, 1980.
[188]   MORET, V. , A. FORSTER, M. C. LQVERRIÉRE, H. LAMBERT, R. C. GAILLARD, P. BOUR-
         GEOIS, A. HAYNAL, M. GEMPERLE and E. BUCHSER: Mechanism of analgesia induced
         by hypnosis and acupuncture: is there a difference? Pain 45: 135-140, 1992.
[189]   MOSKALENKO, YU. YE.: Dynamics of the Brain Blood Volume under Normal Condi-
         tions and Gravitational Stresses. Nauka, Leningrad (Russian), 1967.
[190]   MULDER, P. , E. G. DOMPELING, J. C. VAN SLOCHTEREN- VAN-DER-BOOR, W. D. KUIPERS
         and A. J. SMIT: Transcutaneous electric nerve stimulation (TENS) in Raynauds phe-
         nomenon. Angiology 42: 414-417, 1991.
[191]   MUSAEV, A. V. and F. A. ABDULLAEVA: The neurophysiological and neurochemical
         mechanism of the analgesic action of transcutaneous electrostimulation. Vopr. Kuror-
         tol. Fizioter. Lech. Fiz. Kult. 1991: 69-72 (Russian).
[192]   NASHOLD, B. S., jr.: Dorsal Column Stimulation for Control of Pain; A Three-Year
         Follow-Up. Surg. Neurol. 4: 146-147, 1975.
[193]   NASHOLD, B. S., jr., FRIEDMAN, H.: Dorsal column stimulation for control of pain.
         Preliminary report on 30 patients. J. Neurosurg. 36: 590-597, 1972.
[194]   NICHOLAS, A. D. G., R. H. TIPTON, C. J. WHEATLY and J. BIRCUMSHAW: Obstetric
         Practice and Epidural Analgesia.The J. Obst. Gynec. Brit. Comm. 77: 457-61,
         1970.
[195]   NIELSON, K. D., J. E. ADAMS and Y. HOSOBUCHI: Experience with Dorsal Column
         Stimulation for Relief of Chronic Intractable Pain: 1968-1973. Surg. Neurol. 4:
         148-152, 1975.
[196]   NIV, D, , S. BEN ARI, A. RAPPAPORT, S. GOLDAFSKI, M. CHAYEN and E. GELLER: Post-
         herpetic neuralgia: Clinical experience with a conservative treatment. Clin. J. Pain
         5: 275-300, 1989.

[197] NOLAN, M. F. : Conductive differences in electrodes used with transcutaneous electric nerve stimulation devices. Phys. Ter. **71**: 746-751, 1991.

[198] NYBERG, F., B. G. L. ALMAY, F. JOHANSSON, L. V. KNORRING, T. L. YAKSH and L. TERENIUS: Opioid Peptides and Substance P in Cerebrospinal Fluid – Increased Concentration after High Frequency Transcutaneous Electric Nerve Stimulation. Schmerz-Pain-Douleur, **9** (3a): 165-174, 1988.

[199] PICAZA, J. A,, B. W. CANNON, S. E. HUNTER, A. S. BOYD, J. GUMA and D. MAURER: Pain Suppression by Peripheral Nerve Stimulation, Part I. Observations with transcutaneous Stimuli; Part II. Observations with Implanted Devices. Surg. Neurol. **4**: 105- 128, 1975.

[200] PERT, C. B. and S. H. SNYDER: Opiate Receptor: Demonstration in Nervous Tissue. Science, **179**: 1011-1014, 1973.

[201] PINEDA, A.: Dorsal Column Stimulation and its Prospects. Surg. Neurol. **4**: 157-163, 1975.

[202] PRAGER, P.: Der Wert konventioneller Diagnostik bei artikulären und periartikulären Schmerzzuständen. Therapiewoche **33**: 3825-3832, 1983.

[203] PRICE, D. D. , S. LONG and C. HUITT: Sensory testing of pathophysiological mechanisms of pain in patients with reflex sympathetic dystrophy. Pain **47**: 163-173, 1992.

[204] RAY, C. D,: Conclusions, Reports and Summary of the Minneapolis Pain Seminar on Electrical Stimulation of the Human Nervous System for the Control of Pain. Surg. Neurol. **4**: 202-204, 1975.

[205] RAY, C. D. and D. D. MAURER: Electrical Neurological Stimulation Systems. A Review of Contemporary Methodology. Surg. Neurol. **4**: 82-90, l975,

[206] RICHARDSON, D. E.: Analgesia Produced by Stimulation of Various Sites in the Human Beta-Endorphin System. Applied Neurophysiol. **45**: 116-122, 1982.

[207] RICHARDSON, D. E.: Long-term Follow-up of Deep Brain Stimulation for Relief of Chronic Pain in the Human. Modern Neurosurgery 1 (M. Brock, ed) Springer Verlag, Berlin-Heidelberg: 449-453, 1982.

[208] RICHTER, H.: Die Migräne. In: Handbuch f. Neurol. (Bumke-Förster ed.) **XVII**: 166-245, Springer, Berlin, 1935 .

[209] RODRIGUEZ, E. , M. J. MOIZOISO, M. GARABAL, M. P. FERNANDEZ, L. RODRIGUEZ-BUJAN and A. BELMONTE: Effects of transcutaneous nerve stimulation on the plasma and CSF-concentrations of beta-endorphin and the plasma concentrations of ACTH, cortisol and prolactin in hysterctomized women with postoperative pain. Rev. Espan. Anaestesiol. Reanim. **39**: 6-9, 1992.

[210] SATTER, P. und B. KUNKEL: Methoden der Therapie des Brustschmerzes – Chirurgische Therapie akuter kardiovaskulärer Erkrankugen. Therapiewoche **32**: 2475-2481, 1982.

[211] SAVAGES, N.: Elektroanalgesie beim Zahnarzt. Biomed **3**: 61, l983.

[212] SCHMIDT, R. F.: Neurophysiologie. Heidelberger Taschenbücher Band 96. Springer Verlag, Berlin-Heidelberg-New York, 1971.

[213] SCHNEIDER, F.: Über die vasomotorische Benervung der Extremitäten. Naunyn-Schmiedebergs Arch. exp. Path. Pharmak. **176**: 111-140, 1934.

[214] SCHÖLMERICH, P.: Die Differentialdiagnose des kardiogenen Brustschmerzes. Therapiewoche **32**: 2413-2419, 1982.

[215] SCRIBONIUS LARGUS DESIGNATUS: De compositione medicamentum liber. Ed. J. M. Berthold. Argentor, 1786.

[216] SEYLE, H.: Einführung in die Lehre vom Adaptionssyndrom. G. Thieme-Verlag, Stuttgart, 1953. Übersetzung aus dem Englischen von H. Köbcke. Originaltitel: The Story of the Adaptation Syndrome. Acta Inc. Medical Publishers, Montreal, Canada, 1952.

[217] SHEALY, C. N.: Physiological Substrate of Pain. Headache 6: l0l, 1966.

[218] SHEALY, C. N.: Dorsal Column Electrohypalgesia. Headache 9: 99-108, l969.

[219] SHEALY, C. N.: Dorsal Stimulation: Optimization of Application. Surg. Neurol. **4**: 142-l45, 1975.

[220] SHEALY, C. N., N. TASLITZ, J. T. MORTIMER and D. P. BECKER: Electric Inhibition of Pain: Experimental Evaluation. Anesth. Analg. Curr. Res. **46**: 299-305, 1967.

[221] SHELDEN, C. H, F. PAUL, D. B. JAQUES and R. H. PUDENZ: Electrical Stimulation of the Nervous System. Surg. Neurol. **4**: 127-l32, 1975.

[222] SIEGFRIED, J.: Chronische therapieresistente Schmerzen: Stellenwert und Möglichkeiten der neurochirurgischen Neurostimulationsmethoden. Biomed **4**: 18-24, 1984.

[223] SIMONS, D. G. and J. G. TRAVELL: Myofascial Origins of Low Back Pain. Postgraduate Medicine **73**: 66-108, 1983.

[224] SJOLUND, B. H., L. TERENIUS and M. ERIKSSON: Increased Cerebrospinal Fluid Levels of Endorphins After Electroacupuncture. Acta Physiol. Scand., **100**: 382-384, 1977.

[225] SKINHOJ, E.: Studies on Sympathetic Control of Cerebral Circulation. Acta neurol. Scand. Suppl. **48**: 495, 1971.

[226] SPENCE, A. A. and G. SMITH: Postoperative Analgesia and Lung Function. A Comparison of Morphine with Extradural Block. Brit. J. Anesth. **43**: 144-148, 1971.

[227] STERNBACH, R. A.: Evaluation of Pain Relief. Surg. Neurol. **4**: l99-201, 1975.

[228] STERNBACH, R. A.: Chronic Pain as a Disease Entity. Triangel Sandoz **20**, Nr. l/2: 27-32, l981.

[229] STEUDE, U.: Percutaneous Electro-Stimulation of the Trigeminal Nerve in Patients with Atypical Trigeminal Neuralgia. Neurochirurgia **21**: 66-69, l978.

[230] STOOKEY, B. and J. RANSOHOFF: Trigeminal Neuralgia. C. C. Thomas, Springfield, Ill, 1959.

[231] STULA, D.: Lumbago und Diskushernien. Konservative und operative Behandlung. Biomed **84**: 24-25, 1984.

[232] SWEET, W. and J. WEPSIC: Stimulation of the Posterior Columns of the Spinal Cord for Pain Control. Surg. Neurol. **4**: 133, 1975.

[233] TERENIUS, L.: Biochemische Schmerzmediatoren. Triangel, Sandoz-Zeitschrift für Medizinische Wissenschaft **20**: 19-26, l981.

[234] THOMALSKE, G.: Methoden der Therapie des Brustschmerzes: Neurochirurgische Therapie. Therapiewoche **32**: 2482-2490, 1982.

[235] TOREBJÖRK, H. E. and R. G. HOLLIN: Responses in Human A- and C-Fibers to Repeated Electric Intradermal Stimulation. J. Neurosurg, Psychiatr. **37**: 653-664, 1974.

[236] TRAVELL, J. G.: and D. G. SIMONS: Myofascial Pain and Dysfunction. The Trigger Point Manual. Williams and Wilkins, Baltimore/London 1983.

[237] TULGAR, M. ,F. MCGLONE, D. BOWSHER and J. B. MILES: Comparative effectiveness of different stimulation modes in relieving pain. Part I. A pilot study. Pain **47**: 151-155,1991.

[238] TULGAR, M. ,F. MCGLONE, D. BOWSHER and J. B. MILES: Comparative effectiveness of different stimulation modes in relieving pain. Part II. A double blind controlled long term clinical trial. Pain **47**: 157-162, 1991.

[239] ULANSKI, B.: Conservative Method of Treatment of Trigeminal Neuralgia; Archives of Physical Therapy, X-Ray, Radium 1937: (June 3 pages 5 numbering): A Modified Rapid Sinusosidal Curent for Trigeminal Neuralgia. Archives of Physical Therapy 1940 (May: 4 pages).

[240] ULMER, W. T.: Operative Möglichkeiten des Asthma bronchiale. Prax. Klin. Pneumol. **41**: 577-580, 1987.

[241] URBAN, B. and B. NASHOLD.: Percutaneous Epidural Stimulation of Spinal Cord for Relief of Pain. J. Neurosurg. **48**: 323-328, 1978.

[242] VALDECAS, F. G.: Die Physiologie der Skelettmuskeldurchblutung. Z. Biol. **96**: 28-34, 1935.

[243] WAKIM, K. G.: Influence of frequency of muscle stimulation on Circulation in the stimulated extremity. Arch. Physiol. Med. **34**: 521, 1935.

[244] WALL, P. D.: Excitability Changes in Afferent Fibre Terminations and Their Relation to Slow Potentials. J. Physiol. (London) **142**: 1-21, 1958.

[245] WALL, P. D. and J. R. CROULY-DILLON: Pain, Itch, Vibration. Arch. Neurol. **2**: 365-375, 1960.

227

[246]   WARRICK, J. W.: Stellate Ganglion Block in the Treatment of Menière's Disease and in the Symptomatic Relief of Tinnitus. Brit. J. Anaesth. 41: 699, 1969.

[247]   WELFLING, J.: Der Schulterschmerz – Die Schultersteife. I. Anatomie und Physiologie des Schultergelenks; II. Symptomatologie der Erkrankungen des Schultergelenkes; III. Pathologie des subakromialen Gleitwegs. Folia rheumatologica Geigy, 19a: 3-22; 19b: 23-36; 19c: 39-50, 1969.

[248]   WHITACRE, M. M.: The effect of transcutaneous electric nerve stimulation on ocular pain. Ophthalmic Surg. 22: 462-464, 1991.

[249]   WHITE, L.: Personal Communication

[250]   WOLF, S. L.: Neurophysiologic Mechanisms in Pain Modulation: Relevance to T.E.N.S In: Clinical Transcutaneous Electrical Nerve Stimulation. J. S. Mannheimer and G. N. Lampe, eds., F. A. Davis, Philadelphia, 41-55, 1984.

[251]   WULF, H. , C. MAIER und H. A. SCHELE: Die Behandlung von Zoster-Neuralgien. Anaesthesist 40: 523-529, 1991.

[252]   ZIMMERMANN, M.: Contribution by Thin Myelinated (Group III) Cutaneous Afferent Fibres to Central Nervous Activity as Revealed by Selective Stimulation. J. Phsiol. (London). 224: 33, 1972.

[253]   ZIMMERMANN, M.: Physiologische Mechanismen von Schmerz und Schmerztherapie. Triangel Sandoz-Zeitschrift für medizinische Wissenschaft 20: 7-18. 1981.

[254]   ZIMMERMANN, M.: Periphere Mechanismen von Schmerz und Schmerztherapie. Sandorama (Sandoz AG, Basel): 21-26, 1982.

[255]   ZIMMERMANN, M.: Physiologische Mechanismen des Schmerzes. Therapiewoche 33: 3781-3786, 1983.

[256]   ZIMMERMANN, M. und H. SEEMANN: Der Schmerz – ein vernachlässigtes Gebiet der Medizin? Springer, Berlin, Heidelberg, New York, London, Paris, Tokio, 1986.

This is an excerpt of the relevant literature.

SUBJECT INDEX

Sources of
Figures and Tables:

Fig.	Ref.	Fig.	Ref.
1	254	49	118
2	254	51	128
3	254	52	128
4	139	53	118
5	110	54	118
6	115	57	118
7	115	58	118
8	118	62	118
9	118	63	118
10	116	69a	118
11	118	71	118
11a	120	72	118
pg 35	122	74	118
13	118	76	118
14	107	80	118
15	107	81	118
16	107	82a	118
17	107	103	65
18	118	105	124
19	118	106	124
20	110		
21	110		
22	110	Tabl.	Ref.
25	118		
26	7	1	63
28	120	2	118
31	120	2a	42
32	120	3a	120
33	127	4	118
34	118	5	118
37	118	6	118
39	118	7	7
40	118	pg 87	123
42	118	D-1	119
46	118	D-2	119